JOURNAL OF
Extension

Spring 2022
Volume 60, Issue 2

CLEMSON
UNIVERSITY
PRESS

ISSN 1077-5315

Published by Clemson University Press in Clemson, South Carolina

The *Journal of Extension* is a diamond open-access publication and does not assess article-processing fees of any kind. It is openly available to read online at www.joe.org.

CONTENTS

Feature Articles

Research in Brief

Ideas at Work

Tools of the Trade

JOURNAL OF
Extension

Feature Article

Volume 60, Issue 2, 2022

A Review of Youth Mental Health Curricula in Peer-Reviewed Studies Addressing Access, Equity, and Belonging

MONICA M. LOBENSTEIN[1], JENNIFER PARK-MROCH[1], LANA LICHFIELD CROWLEY[2], COLEY BEAN[3], AND MAREN WRIGHT[4]

AUTHORS: [1]University of Wisconsin-Madison Division of Extension. [2]University of Utah. [3]Utah State University Extension. [4]Utah State University.

Abstract. The goal of this literature review was to identify evidence-based curricula that support youth mental health with special attention to inclusion of access, equity, and belonging (AEB). Four databases were searched for peer-reviewed articles published between 2010 and 2019 related to youth mental health curricula. A total of 1446 articles were identified, and 171 articles underwent a full-text review. Of the 61 curricula identified, 44% addressed AEB to some extent and 65% showed program effectiveness. Four programs were recommended (Sources of Strength, Teen Mental Health First Aid, Dynamic Mindfulness, and Youth Mental Health First Aid) and eight conditionally recommended.

INTRODUCTION

Worldwide, half of all mental illnesses begin before the age of 14 and three-quarters begin before young adults reach their mid-20s (Kessler et al., 2007). In the United States, two of the most diagnosed mental health disorders are anxiety and depression and are often found to be co-occurring (Ghandour et al., 2019). Promotion of mental health in young people occurs through early education and skills building and addressing and preventing mental health issues (Patel et al., 2007). Many curricula and educational resources exist that address mental health and can be promoted among school-age youth.

4-H and other Extension youth programs are built on the conceptual framework of positive youth development (PYD), which is well positioned to implement appropriate curricula in community-based settings that complement the efforts of schools. In Extension and for the purposes of this paper, PYD is defined as an intentional, prosocial approach to developing thriving youth (National Institute on Food and Agriculture [NIFA], 2016). With its roots in prevention, the focus of PYD creates a shift from deficit-based to asset-based strategies and life skills development, implemented most often in the community-based contexts in which youth live.

A vital part of effectively implementing educational programs is understanding community contexts, detailing subtle differences in the intent and development of curricula, and understanding how the curricula interact with those (Smith et al., 2017; Vance, 2012). PYD approaches, such as encouraging personal agency and respecting youth, have been shown to result in greater resiliency and sense of well-being in youth (Sanders et al., 2015). This PYD lens is an integral part of how Extension educators anticipate using curricula that support mental health in young people.

Considering a youth's context is part of the PYD approach. Youth from minoritized or underserved populations experience additional risk factors including historical trauma, alienation, acculturation, racism, and discrimination. Unfortunately, injustice is evident in Extension's past programming. Historically, Black individuals have experienced discrimination in the Cooperative Extension Service through unequal racial policies, particularly in the southern branches of Extension. Black leadership protested the racial policies from the beginning and worked for change. Resistance to these efforts was notable, as one Extension official expressed, "the Extension Service is not an integration agency. We are the education agency of [the] USDA for Agriculture, Home Economics, and Youth and related programs" (Harris, 2008). This racial inequality eventually led to the disappearance of a Black Extension force and years of lost representation (Harris, 2008). In addition to racial inequality, approaches to relationship education and marriage advice in 4-H programming have left little room for LGBTQ+ youth (Rosenberg, 2015).

Improving Extension's outreach to minoritized youth with an inclusive approach to PYD is important, especially when looking at the evidence of the increased risks experienced by youth and the intersectionality of mental health and minoritization. For example, between 1991 and 2017, the suicide rate among Black youth increased more than 70% (Lindsey et al., 2019). The suicide rate for Indigenous young people is more than double the rate of non-Hispanic white youth (National Center for Injury Prevention and Control, 2019). Youth who identify as lesbian, gay, bisexual, transgender, queer, or other minority gender and sexual identities are at greater risk for depression, anxiety, substance abuse, and suicidality than their gender-conforming or heterosexual peers (Russell & Fish, 2016).

In the context of access, equity, and belonging (AEB), the following definitions guided this paper. Access refers to the opportunities youth have to engage in services, programs, and experiences. It is the result of individual circumstances; structural or systemic barriers and supports; design and availability of a service, program, or experience; or some combination of these (Osher et al., 2020). Expanding access is the intentional multi-layered work of systems and design changes that remove barriers and increase opportunity (Osher et al., 2020). Equity is "when a person or group receives the unique resources and opportunities needed to reduce or eliminate the barriers" (Fields, 2019). As an essential element of 4-H, belonging can be defined as when "youth feel like they belong in a program that is safe and a positive environment" (Kress, 2005). Inclusion is a deeper function of belonging related to equity and "is the act of creating a space where each person is authentically valued, respected and supported" (Fields, 2019). Peer-reviewed studies of curricula with embedded stigma reduction strategies and culturally competent practices that build on existing protective factors can help address AEB as youth learn about concepts and skills that promote mental health (Siegel et al., 2011).

The need for a systematic review was identified by the Mental Health and Well-Being (MHWB) Champion Group, a subcommittee of 4-H's Access, Equity, and Belonging Committee (AEBC). The purpose of the AEBC (2020) and its subcommittees was to increase the capacity of the 4-H system to achieve its vision to "reflect the population demographics, vulnerable populations, diverse needs and social conditions of the country" (4-H Program Leaders Working Group, 2020). A part of the group's formal charge was to recommend curricula for use by Extension educators, with each curriculum being appraised for how it addressed AEB among diverse youth and helped prevent mental health crises. Thus, the primary objective of this systematic review was to identify available evidence-based curricula that generally promote the mental health and well-being of school-age youth and have proven effectiveness, as shown through multiple peer-reviewed articles. The results are used

to make recommendations to Extension educators about available and effective programs suitable for the community-based contexts in which they work. The results will also assist in identifying gaps in what is available for those contexts and determining how Extension educators may help fill those gaps.

METHODS

SEARCH STRATEGY

Four electronic reference databases (ERIC, PsycINFO, PubMed, Web of Science—Social Science Citations Index) were searched on August 15, 2019, within the Abstract and Title fields. The search terms were broad with the intent of capturing all potentially relevant studies. An information specialist and reference librarian were consulted to identify the most appropriate search terms to address the above stated goals of the literature review. The search terms were (((youth or teen* or adolescen* or juvenile*) AND ((mental or emotion* or behav*) and (health* or wellness or well-being or wellbeing)) in Title AND (curricul* or program*) in Abstract or Title). For all systematic or literature review articles identified in the original search, a manual search of the reference lists of review articles was conducted to discover additional papers meeting the search criteria that were not identified in the original database search.

ELIGIBILITY CRITERIA

In addition to using the search terms above, the inclusion criteria for the search were: articles published from January of 2010 to August of 2019; peer-reviewed journals; and programming conducted on a human population group. Limiting the search to the most recent decade of articles was selected to allow for studies that address current contexts of youth mental health and societal considerations of AEB.

Subject categories were adolescents or children (PsycINFO—ages 13-17 and 6-12) and mental/emotional/behavioral health. The search was limited to articles published in English. Titles and abstracts were examined by two reviewers for further exclusion. The exclusion criteria were three broad categories: 1) not related to youth; 2) not related to mental health; and 3) not related to curriculum or education. Abstracts included must have held an indication that the curriculum was used to benefit youth mental health, which was defined as school-age (generally ages 5-18, though any application in this age range was acceptable). Abstracts must have included an indication that the curriculum had a focus on mental or behavioral health, not simply a mention of behavioral health as a measure of outcomes. The abstract must have included a description of the use of a standardized, defined curriculum that would be available for general use, either purchased or shared.

SEARCH OUTCOMES

Using the search terms, 1446 records were found and 276 duplicates were removed. Two researchers independently reviewed the contents of the titles and met to discuss discrepancies in inclusion criteria; there was disagreement on 138 titles, largely related to identifying curricula inclusion. If either researcher identified the article title as meeting inclusion criteria, the article was retained. From the initial review, 838 articles were excluded (39 not related to youth; 99 not related to mental health; 699 not related to education or curriculum; one not in English). There were 332 titles remaining at the abstract review stage, and inclusion and exclusion criteria remained unchanged. An additional 215 articles were excluded (10 not related to youth; 13 not related to mental health; 192 not related to curriculum), leaving 117 articles remaining for full text review.

Of the 117 articles, 29 were deemed a systematic or literature review and were separated for an additional step of a reference review, with the reference lists culled for articles that would meet the original inclusion/exclusion criteria by title. The review of reference lists added an additional 126 articles meeting criteria by title. Of these articles, 43 were removed by abstract review, applying the same three exclusion criteria (not related to youth, mental health, or curricula). A total of 88 articles from this reference review, which were identified as meeting inclusion criteria by title and abstract and were included in the full-text article review, were combined with the 83 from the original search, leaving a total of 171 articles which underwent a full-text review (see supplementary material for complete list).

The following information was extracted from each full-text article and was reported in a predesigned table: authors; first author last name; publication year; article title; journal title; name of program; article type; article aims; study location; sample size; sample type; curriculum setting; curriculum age group; demographic focus; tier of prevention; curriculum delivery person; curriculum format; curriculum focus; curriculum number of sessions; curriculum length of sessions; curriculum training costs; curriculum implementation costs; curriculum access; program effectiveness; and addresses AEB.

THEMATIC ANALYSIS AND PROGRAM RECOMMENDATIONS

A thematic review of the 171 full-text articles was conducted and focused on curricular design elements as well as research study design elements in the identified articles. Curricular design elements emphasizing the AEB theme included stigma reduction, cultural competence, empathy with others, relationships/friendship, social skills, or peer support. Study design elements included in the review used culturally competent practices to serve diverse racial, cultural, sexual, gender, socio-economic status, or other identities.

Program recommendations were developed to efficiently summarize the study findings in a manner that supports program implementation by Extension educators. The first criterion was sufficient peer-reviewed evidence to support implementation as an evidence-based program (EBP), measured by having at least three peer-reviewed studies in the literature. All programs without three peer-reviewed studies were reviewed against federally funded lists of EBPs; those programs were included. The second criterion was whether the program could be implemented in non-clinical settings by non-clinical facilitators, that is, Extension educators.

RESULTS

Of the 171 studies reviewed, 70% were quantitative in nature, 10% were qualitative, 10% were mixed methods, and 10% presented pre-outcome or methods were unspecified. Study types were reviewed and showed that 33% had an experimental design, while 5% used a quasi-experimental design and 47% were non-experimental with convenience sampling. Roughly one-third of the studies had sample sizes of 100 or fewer, one-third ranged from 101 to 500, and another third were over 500. Fewer than half of the studies were conducted in the United States. Most of the studies (65%) demonstrated effectiveness in the hypothesized outcome.

In the final analysis, 75 of the 171 studies reviewed addressed AEB in at least a small way. The majority (62%) of the study methods used classroom-based or in-school instruction, with the remaining spread across clinical, after-school, home, community, online, or professional settings. In addition, the study methods specified a universal prevention approach in 52% of the studies and a targeted prevention approach in 38%.

In all, 60% of the curricula focused on youth mental health as a general or non-specific category. An additional 7% explicitly focused on suicide prevention, one study focused on self-harm, and 10% were tailored to anxiety and depression management. Taken together, three-fourths of the studies specifically utilized curricula and methods to directly address youth mental health and harm prevention. Building protective factors related to skill-building in resilience, mindfulness, positivity, and life skills was the focus of 23% of the curricula.

The larger the sample size and the more methodologically rigorous the study, the less likely it was to address AEB. Of studies using experimental, randomized, and control methods, 72% had positive findings, yet only 25% addressed AEB. Of the quasi-experimental designs, 38% addressed AEB, and in studies using convenience or non-experimental methods, 47% addressed AEB. A similar pattern was identified based on study sample size. In studies with 100 or fewer participants, 46% addressed AEB. With sample sizes of 101-500, 57% addressed AEB. Yet, with larger studies of sample sizes over 500, only 35% addressed AEB.

PROGRAM RECOMMENDATIONS

This review of 171 studies identified 61 distinct curricula that aimed to address youth mental health. No individual program in our review had more than four studies to support its effectiveness, and there was insufficient replication and quantity of data to make firm recommendations about the utility of the AEB-focused curricula. Recommendations also took into consideration if the program could be implemented by a non-clinical professional, the training and participant costs, and the manualization or ease of implementation by Extension educators. Though each identified curriculum demonstrated effectiveness in supporting youth mental health, each only minimally addressed AEB. Complete results are found in the Appendix.

Four programs were identified that either had a minimum of three peer-reviewed articles that supported the program's effectiveness or were listed by a federally funded agency as an EBP. These programs include Sources of Strength, Teen Mental Health First Aid, Dynamic Mindfulness, and Youth Mental Health First Aid. Sources of Strength had effective use of a social network model that identified and trained "natural leaders" to form inclusive peer networks around positive social norms. Teen Mental Health First Aid (T-MHFA) emphasized peer support and reducing stigma. The training and curricula for Dynamic Mindfulness, formerly known as Transformative Life Skills, had an entire unit that focused on healthy relationships and some belongingness and inclusion aspects. Youth Mental Health First Aid (Y-MHFA) is like T-MHFA with manualized implementation but focuses on training adults who work with youth.

Eight programs were conditionally recommended, as they were determined to be either more difficult for educators to access and implement or having a narrower evidence base. The Adolescent Depression Awareness Program (ADAP) is conditionally recommended, with studies showing initial effectiveness, supported by Johns Hopkins. The path to training was not clearly laid out, and it is labeled as school-based. However, this program shows promise and could be a good fit for community implementation. Creating Opportunities for Personal Empowerment (COPE); Yellow Ribbon; and Question, Persuade, Refer (QPR) are available and affordable, but effectiveness is in early stages or mixed. Zippy's FRIENDS and Youth Aware of Mental Health are widely used internationally with a strong evidence base but are less available to educators within the United States. In Your Own Voice recommends that trainers be individuals in recovery. Positive Action is intended to be implemented school-wide or district-wide, led by teachers at each level.

The University of Montana Extension is currently playing an integral role in piloting Youth Aware of Mental Health programs in the United States, and future review will determine if this becomes a recommended program.

Several resources that emerged during the review—including Big Brothers Big Sisters, Prodigy, and Turn 2 Us—were existing programs but did not have a defined curriculum for implementation. Further, two of these are individual, localized programs which do not appear to have been replicated elsewhere. Though much can be learned from the methods of these three programs, they are not yet applicable for Extension educators. Additional information or resources for five programs (Aussie Optimism Program, Project Wings, FRIENDS, Skills for Life, and Making the Link) were extremely challenging to find, and with limited or no availability are not recommended for use by Extension educators. Others, like Bounce Back, CLIMB, and MBSR were designed to be implemented by clinicians or highly trained professionals, making implementation in Extension programs unlikely. The Family Bereavement Project was too narrow in scope to be recommended.

FUTURE RECOMMENDATIONS AND CONCLUSION

In all, findings indicated that the larger and more rigorous studies were less likely to specifically address AEB. It is likely that AEB considerations in programming to address youth mental health are still in their infancy and need more time to come to the forefront. As further program development and research on their effectiveness continues, it is critical that AEB be a distinct focus, there be standardization for ease of replication, and the proliferation of curricula be avoided. Only if these things occur can a strong evidence-base be established for curricula that incorporate AEB and support youth mental health. Additionally, it will be particularly important for larger, well-funded studies to include consideration of AEB in the methodological approach and to rigorously evaluate the utility and applications of these constructs in youth mental health curricula.

Given this foundational understanding of existing curricula and their effectiveness, Extension is well-positioned to work toward identifying gaps related to AEB in mental health curricula. Within those gaps, educators may also play a unique and integral role in identifying, adapting, or developing strategies to help create more AEB in community-based mental health education programs.

REFERENCES

4-H Program Leaders Working Group. (2020, July 17) *Access, Equity, and Belonging Committee*. https://access-equity-belonging.extension.org/

Fields, N.I. (2020). *Literature Review: Exploring 4-H Thriving through an Equity Lens.* Extension Committee on Organization and Policy. https://access-equity-belonging.extension.org/wp-content/uploads/2020/02/Literature-Review-Thriving-Model-Equity-Lens-FINAL-4.pdf

Ghandour, R. M., Sherman, L. J., Vladutiu, C. J., Ali, M. M., Lynch, S. E., Bitsko, R. H., & Blumberg, S. J. (2019). Prevalence and treatment of depression, anxiety, and conduct problems in us children. *The Journal of Pediatrics, 206*(3) 256-267. https://doi.org/10.1016/j.jpeds.2018.09.021

Harris, C. V. (2008). "The Extension Service Is Not an Integration Agency": The Idea of Race in the Cooperative Extension Service. *Agricultural History, 82*(2), 193-219. doi.org/10.3098/ah.2008.82.2.193

Kessler, R. C., Amminger, G. P., Aguilar-Gaxiola, S., Alonso, J., Lee, S., & Ustun, T. B. (2007). Age of onset of mental disorders: A review of recent literature. *Current Opinion in Psychiatry, 20*(4), 359–364. https://doi.org/10.1097/yco.0b013e32816ebc8c

Kress, C. (2005). *Essential Elements of 4-H youth development.* Chevy Chase, MD: National 4-H Headquarters, CSREES USDA.

Lindsey, M. A., Sheftall, A. H., Xiao, Y., & Joe, S. (2019). Trends of Suicidal behaviors among high school students in the United States: 1991–2017. *Pediatrics, 144*(5). https://doi.org/10.1542/peds.2019-1187

National Center for Injury Prevention and Control. (2019). WISQARS — Web-based Injury Statistics Query and Reporting System. Centers for Disease Control and Prevention. https://www.cdc.gov/injury/wisqars/index.html

National Institute on Food and Agriculture (2016). *Fact sheet: The science of positive youth development.* https://nifa.usda.gov/sites/default/files/resources/Science%20of%20Positive%20Youth%20Development.pdf

Osher, D., Pittman, K., Young, J., Smith, H., Moroney, D., & Irby, M. (2020). *Thriving, robust equity, and transformative learning & development: A more powerful conceptualization of the contributors to youth success.* American Institutes for Research and Forum for Youth Investment. https://www.air.org/sites/default/files/Thriving-Robust-Equity,-and-Transformative-Learning-and-Development-July-2020.pdf

Patel, V., Flisher, A. J., Hetrick, S., & Mcgorry, P. (2007). Mental health of young people: A global public-health challenge. *The Lancet, 369*(9569), 1302-1313. https://doi.org/10.1016/s0140-6736(07)60368-7

Rosenberg, G. N. (2015). *The 4-H harvest: Sexuality and the state in rural America.* University of Pennsylvania Press.

Russell, S. T. & Fish, J. N. (2016). Mental health in lesbian, gay, bisexual, and transgender (LGBT) youth. *Annual Review of Clinical Psychology, 12*, 465-487. https://doi.org/10.1146/annurev-clinpsy-021815-093153

Sanders, J., Munford, R., Thimasarn-Anwar, T., Liebenberg, L., & Ungar, M. (2015). The role of positive youth development practices in building resilience and enhancing well-being for at-risk youth. *Child Abuse & Neglect, 42*, 40–53. https://doi.org/10.1016/j.chiabu.2015.02.006

Siegel, C., Haugland, G., Reid-Rose, L., & Hopper, K. (2011). Components of cultural competence in three mental health programs. *Psychiatric Services, 62*(6), 626-631. https://doi.org/10.1176/ps.62.6.pss6206_0626

Smith, M. H., Worker, S. M., Meehan, C. L., Schmitt-Mc-Quitty, L., Ambrose, A., Brian, K., & Schoenfelder, E. (2017). Defining and developing curricula in the context of Cooperative Extension. *Journal of Extension, 55*(2). https://tigerprints.clemson.edu/joe/vol55/iss2/22/

Suicide Prevention Resource Center. (2013). Risk and Protective Factors in Racial/Ethnic Populations in the U.S. Retrieved August 11, 2020, from http://www.sprc.org/resources-programs/risk-protective-factors-racial-ethnic-populations-us

Vance, F. (2012). An emerging model of knowledge for youth development professionals. *Journal of Youth Development, 7*(1). https://doi.org/10.5195/jyd.2012.151

APPENDIX. PROGRAMS FROM PEER-REVIEWED STUDIES REVIEWED FOR RECOMMENDATIONS

Program Title and Location	Program Description	Facilitator Training	Access, Equity, and Belonging Recommendation and Rationale
Sources of Strength— https://sourcesof strength.org/	A best practice youth suicide prevention project designed to harness the power of peer social networks to change unhealthy norms and culture, ultimately preventing suicide, bullying, and substance abuse. **Audience:** Youth in middle or high school **Setting:** School-based **Time:** 3–4-hour training for peer leaders; programming takes about 40 hours over 3–6 months; designed to be a multi-year program	Adult Advisor training required 3–6-hour orientation training plus monthly support calls **Cost:** Base $5000/school (Additional services available)	Recommended Recommended as evidence-based by HRSA's Rural Health Information Hub. Implements a social network approach to spread positive social change. Peer leaders identified as "natural leaders" representing "diverse social groups."
Teen Mental Health First Aid (T-MHFA)— https://www.mental healthfirstaid.org/ population- focused- modules/teens/	T-MHFA is an evidence-based training that teaches high school students in the 10th, 11th, and 12th grades the skills they need to recognize and help their friends with mental health and substance use challenges and crises and how to get the help of an adult quickly. **Audience:** Youth in grades 10–12 **Setting:** School or community **Time:** 3, 90-minute interactive classroom sessions or 6, 45-minute sessions, on non-consecutive days	Certification required 3-day interactive facilitator training **Cost:** $3,200/person (Group rates for host sites are available.)	Recommended Adult version designated as evidence-based by National Registry for Evidence-Based Programs. Studies show that the program increases youth supportive behaviors with peers and reduces stigma.
Dynamic Mindfulness (DM)—https:// dmind.org/ curriculum/	DM, formerly known as Transformative Life Skills, is a strengths-based curriculum intervention, which includes three core practices (ABCs): yoga postures (Action), breathing techniques (Breathing), and centering meditation (Centering). Within these practices, youth learn cognitive-behavioral therapy strategies to learn skills for self-awareness, impulse control, and managing anxiety and stress. **Audience:** Youth of any age **Setting:** School-based **Time:** 48, 15-minute lessons	Training recommended (not required) 6 hours for Foundations course; 6 hours for Teacher course **Cost:** $35 to purchase curriculum; facilitator training offered on a sliding scale $100–150	Recommended CASEL-designated evidence-based program. One of four units addresses "Healthy Relationships" and studies in a range of diverse school settings show increases in prosocial behaviors and decreases in hostility among participants.
Youth Mental Health First Aid (Y-MHFA)— https://www.mental healthfirstaid.org/ population- focused-modules/ youth/	Mental Health First Aid is an 8-hour course that teaches awareness of mental health problems and crisis and the skills to identify, understand, and respond to signs of addictions and mental illnesses. **Audience:** Adults who work with Youth/Teens and Children **Setting:** Community **Time:** 8-hour training	3–4-day facilitator training **Cost:** $1,500 facilitator trainer costs, manual required for each participant	Recommended Adult version designated as evidence-based by National Registry for Evidence-Based Programs. Studies suggest effectiveness in training adults to support youth mental health, addresses equity, widely available, and manualized.

A Review of Youth Mental Health Curricula in Peer-Reviewed Studies

Program Title and Location	Program Description	Facilitator Training	Access, Equity, and Belonging Recommendation and Rationale
Adolescent Depression Awareness Program (ADAP) — https://www.hopkinsmedicine.org/psychiatry/specialty_areas/moods/ADAP/	The Adolescent Depression Awareness Program (ADAP) educates school-based professionals, high school students, and parents about the illness of depression. **Audience:** Grades 9-12, Professionals, and Community Members **Setting:** School and Community based **Time:** 3 hours, recommended in 2-3 sessions	No information available.	Conditional Recommendation Initial studies show effectiveness and the program addresses stigma reduction as a key objective.
Creating Opportunities for Personal Empowerment (COPE)—https://www.cope2thriveonline.com/	A manualized cognitive-behavioral skills building intervention program that can be delivered in primary care, school-based clinics, and mental health settings **Audience:** Children (ages 7–11); Teens (ages 11–18); Young Adult (ages 18–24) **Setting:** School-based, clinical, or self-paced online **Time:** 7 sessions, 20-50 minutes each depending on delivery style	4-hour online facilitator training **Cost:** $385 for facilitator training, $5 (self-print) to $20 printed manual required for each participant	Conditional Recommendation Initial studies show effectiveness, and the program has been adapted for multicultural audiences.
In Our Own Voice (NAMI)—https://www.nami.org/Support-Education/Mental-Health-Education/NAMI-In-Our-Own-Voice	In Our Own Voice is a presentation that provide a personal perspective of mental health conditions **Audience:** Teens (ages 11–18) and Community Members, Spanish language limited availability **Setting:** Community **Time:** 40-, 60- or 90-minute presentations	No information on Training. **Cost:** Free	Conditional Recommendation The presenters with lived experience are usually from the local NAMI chapter. Young adults should be presenters.
Positive Action— https://www.positiveaction.net/	Positive Action is a modular social and emotional learning program that embeds academic content in lessons designed to develop an intrinsic interest in learning and promote prosocial behavior. **Audience:** Pre-K to 8th Grade **Setting:** School based **Time:** 140 sessions lasting 15–20 min for K–6 (4 days a week) & 70 session lasting 20min for older children (2 days a week)	2-hour online webinar, on-site orientation, or on-site training of trainers available **Cost:** $550 for the online webinar and $3000 for the other options.	Conditional Recommendation Initial studies show effectiveness with low-income and ethnic minority youth. The program requires significant time commitment.

Program Title and Location	Program Description	Facilitator Training	Access, Equity, and Belonging Recommendation and Rationale
Question, Persuade, Refer (QPR)— https://qprinstitute. com/	QPR is intended to reduce suicidal behaviors and save lives by providing innovative, practical, and proven suicide prevention training. This training works to empower all people, regardless of their background, to make a positive difference in the life of someone they know. **Audience:** Adults (adaptable for youth) **Setting:** Flexible for almost any setting **Time:** 1 hour for adults; typically longer for youth audience	Facilitator training required 12-hour facilitator training **Cost:** $495/person for a 3-year certification (renewable)	Conditional Recommendation Studies show evidence of effectiveness with adults, the primary audience for the program, which has been adapted for use in training teen audiences.
Youth Aware of Mental Health (YAM)—http:// www.y-a-m.org/	YAM is a universal mental health intervention intended to raise mental health awareness, enhance skills and emotional resiliency, and empower youth. **Audience:** Youth in grades 7–12 **Setting:** School-based **Time:** 5 weekly, 1-hour lessons	Two trained adults conduct every YAM program. One of these adults must have completed an intensive week-long certification before becoming a YAM instructor and the other one a shorter training to act as support in the classroom.	Conditional Recommendation Emerging research shows that YAM is promising for use in the United States. Program focus is placed on youth voice with emphasis on understanding different perspectives and problem solving.
Yellow Ribbon— https://yellow ribbon.org/	The Light for Life Foundation International/ Yellow Ribbon Suicide Prevention Program® removes barriers to helping and makes suicide prevention accessible through community-based instruction and resources. **Audience:** Youth/Teen and Children **Setting:** Community **Time:** 30–60-minute training	3-hour facilitator training offered online **Cost:** $295 for facilitator training	Conditional Recommendation Studies were mixed on effectiveness and the program AEB aspects touch on more basic concepts such as stigma reduction. Low barriers to implementation suggest potential utility.
Zippy's Friends— https://casel.org/ zippys/	Zippy's Friends is a skills promotion program for social and emotional learning teaching positive problem solving. **Audience:** Ages 5-8 **Setting:** School-based **Time:** 45 minutes, 24 sessions	6-hour facilitator training and a 2-day train-the-trainer option **Cost:** Unknown	Conditional Recommendation Studies suggest effectiveness in international implementation. Difficult accessibility in the United States reduced recommendation to conditional.

JOURNAL OF
Extension

Feature Article

Volume 60, Issue 2, 2022

Extension and Tourism: Responding to the Changing Needs of Society and New Opportunities

Doug Arbogast[1], Daniel Eades[1], Stephan Goetz[2], and Yuxuan Pan[2]

AUTHORS: [1]West Virginia University. [2]Pennsylvania State University.

Abstract. This study highlights the results of a national survey of Extension land-grant and sea grant professionals designed to better understand their involvement in state/regional tourism programming and their perceptions of tourism related opportunities and challenges. This study demonstrates the breadth and importance of Extension's tourism programing and continued challenges including limited investment and commitment by state institutions and the larger CES for core tourism program offerings. Investments in tourism programing are recommended as a way for Extension to maintain its relevancy, and better engage and address the community and economic development needs of traditional and emerging audiences.

INTRODUCTION

Tourism and recreation activities are a rapidly growing segment of the economy and an increasingly important diversification strategy for rural economies. As shown in Figure 1, employment in recreation is both widely dispersed and a significant share of the local economy in many regions of the country (Bureau of Labor Statistics, 2019). In fact, every state contains multiple counties in which the share of employment in leisure and hospitality is greater than 14% of total employment, placing them in the top third of counties by employment share. Over the past two decades, employment in recreation and tourism activities such as Arts, Entertainment, and Recreation (NAICS 71) and Accommodation and Food Services (NAICS 72) has grown by 25% and 32% respectively (Rasker, 2018). Prototype statistics from the Bureau of Economic Analysis's Outdoor Recreation Satellite Account (ORSA) show that the outdoor recreation sector alone employs approximately 4.5 million workers, comparable to all nondurable goods, such as manufacturing (4.6 million jobs) and the nation's hospitals (5.0 million jobs) (Highfill et al., 2018). The value-added contribution of the outdoor recreation industry to gross domestic product ($372 billion, or 2.2%) and overall economic growth continues to increase and represents a larger share of GDP (2.2%) than many industry sectors that are commonly the focus of economic development efforts in rural areas, such as mining (1.2%) and utilities (1.6%). This contribution is also more than double the contribution of U.S. farm production (0.7%) (Highfill et al., 2018; U.S. Bureau of Economic Analysis,

2020). Recent research has also documented that counties with greater recreational sector employment were more resilient to the employment shocks associated with the 2008 Recession (Boettner et al., 2019).

The Cooperative Extension Service's (CES) role in tourism development has been documented by organization leaders since the mid-20th century, and Extension is continually called upon to deliver educational programs that facilitate informed decisions about tourism investments, tourism's contribution to community economic development strategies, and a host of other tourism management, planning, and policy issues. At the National Workshop on Cooperative Extension's Role in Outdoor Recreation in 1967, R.P. Davison, director of the Vermont Extension Service, explicitly identified a need for "recreation and tourism programming...to become an integral part of Extension programs." This call for sound programming that could assist local leaders and governing officials in making informed decisions about recreation and tourism as a component of total community development was reiterated a decade later in the CES Recreation and Tourism Task Force report (1978), and again in 1990 when Beth Walther Honadle, National Program Leader for Economic Development, highlighted the important intersections between Cooperative Extension programming and tourism development, including opportunities for partnerships and interdisciplinary programming and an explicit recognition of the tourism industry and Extension's roles in diversifying and revitalizing rural communities and economies (Honadle, 1990). More recently, National Sea Grant identified tourism as a major driver of economic growth and a key component

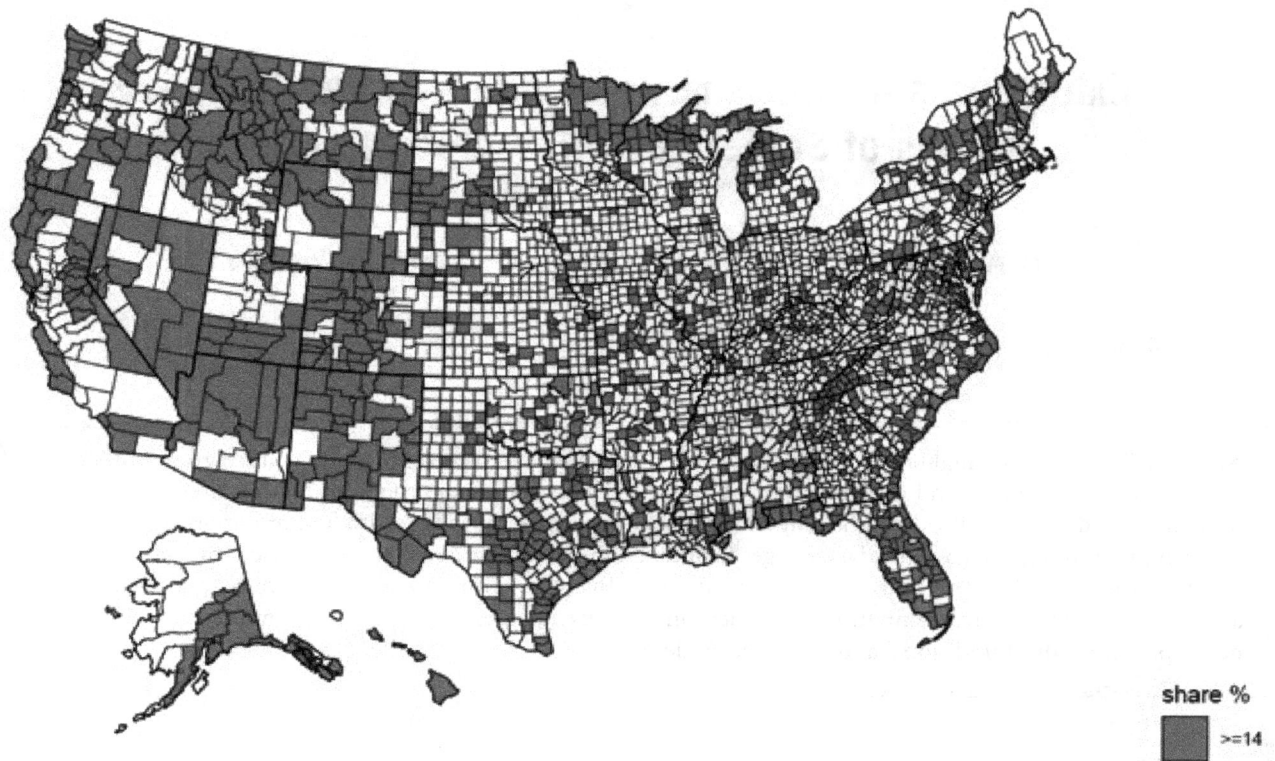

Figure 1. Counties with the largest share of employment in leisure and hospitality. Data are from Bureau of Labor Statistics (2019) Quarterly Census of Employment and Wages [2019 Q3 Employment]. Retrieved from http://data.bls.gov and authors' calculations.

of broader community development goals and defined their role in sustainable coastal tourism activities (NOAA, 2018).

Despite acknowledging the industry's importance to constituents, the need for Extension to respond to the changing needs of society, and Extension's role in providing sound programming to facilitate its development, CES has made few investments to facilitate the development and dissemination of successful tourism programming. The National Extension Travel and Tourism Advisory Committee (1993) described the CES's tourism efforts as "lacking a central focus, being unevenly distributed among the States and offering only sporadic opportunities to share resources." And, while it identified substantial involvement by Extension professionals across the country, it also described efforts that were "not at a level commensurate with the industry's economic importance and growing need for assistance."

The National Extension Tourism Design Team (NET) was formed in 1993 as an outgrowth of these initial reports with a mission to integrate research, education, and outreach within Cooperative Extension and Sea Grant to support sustainable tourism, thus contributing to the long-term economic development, environmental stewardship, and socio-cultural wellbeing of communities and regions. NET includes representatives from the USDA Regional Rural Development

Centers (RRDC), USDA NIFA, and two representatives from Canada. The Design Team is a loosely structured group with two tourism-focused land grant and sea grant representatives from each region. Key external stakeholders for NET are federal agency partners such as USDA, U.S. Forest Service, Environmental Protection Agency, Economic Development Administration, National Park Service, NOAA/National Sea Grant Program, National Travel & Tourism Office, and U.S. Department of Commerce/ International Trade Association, as they most align with Extension and tourism (either via direct reports, funding, research, and/or programs).

NET provides a forum for sharing research and best practices in sustainable community tourism. For example, the Design Team has hosted biennial conferences attracting between 50 and 200 attendees since 1995; recently, NET leadership collaborated with the World Leisure Centre of Excellence in Sustainability and Innovation to produce the book *Innovative and Promising Practices in Sustainable Tourism* to showcase case study research examples of Extension tourism programming. Extension tourism research is further enhanced by cooperative research efforts between Extension field professionals and colleagues at the RRDCs.

Finally, NET continues to engage in strategic activities to elevate the role of Extension tourism. In February 2020, land grant and sea grant representatives met to determine how the Design Team could expand its support to grow and enhance the provision of tourism programming within both land grant and sea grant Extension. One outcome of the retreat was the National Extension Tourism Strategic Plan 2020-2025, which serves as a roadmap for sustainable tourism research, education, and outreach within Cooperative Extension and sea grant in collaboration with land grant universities, federal agencies, and other partners. The document reaffirms the core values of Extension tourism first laid out in 1993 and considers a pathway for elevating Extension tourism activities and broadening their reach, including further engagement of faculty and students at land grant universities for delivering applied scholarship to community stakeholders; inventorying and assessing existing Extension and sea grant tourism programs; providing professional development opportunities for Extension and sea grant staff and other key stakeholders; and sharing applied research, special programs, or other tourism development and outdoor recreation technical expertise with key stakeholders and audiences (Mission, Vision and Strategic Plan, n.d.).

In order for Extension to maintain its relevance, it needs to be prepared and willing to engage a broader constituency and adapt programming to existing needs, and CES as an organization needs to encourage and support these efforts. The goal of this study was to better understand the national context of Extension's tourism programming and resource allocation and provide recommendations for strengthening Extension tourism development initiatives.

METHODS

Over the course of 2017 and 2018, the authors (with support from the National Extension Tourism Design Team) designed and conducted a national survey of Extension land-grant and sea-grant professionals to better understand their involvement in state and regional tourism programming and their perceptions of tourism related opportunities and challenges. The survey instrument included multiple choice and open-ended questions designed to measure time allocated to tourism work, topical areas covered and specific programs offered, gaps in tourism program offerings, and challenges educators face in being able to deliver or support tourism.

Data were collected through structured questionnaires administered online using Qualtrics survey software. The samples were drawn from the NET contact list, Northeast Regional Rural Development Center contact list, and National Sea Grant contact list and posted on the National Association of Community Development Extension Professionals Facebook page using the snowballing

technique for dissemination. Data analysis included descriptive techniques to establish general trends.

A total of 133 surveys were completed, including 96 from land-grant Extension, 25 from sea-grant Extension, and 12 responses from individuals not directly affiliated with an Extension Service (Farm Bureau, consultant, entrepreneur, state government, academic faculty, non-profit) (Appendix B). The response rate is unknown. Appendix B shows the number of survey responses according to state, region, and affiliation. Responses were received from 39 states. Nonresponding states included Arizona, Illinois, Kentucky, Missouri, South Dakota, Utah, Washington DC, and Wyoming. The US territories of Guam, Micronesia, Puerto Rico, Samoa, and the US Virgin Islands are included in the RRDC regions, but surveys were not distributed to those territories. Multiple responses were received from several states including 21 from Pennsylvania (16% of responses), 14 from Alabama (10%), nine from Ohio (7%), and five from West Virginia (4%).

RESULTS

Most survey respondents (91% land grant and 60% sea grant) reported having a full-time Extension appointment; however, few respondents reported focusing a large percentage (>80%) of their time on tourism related projects. Only 6% of respondents from land grant and 8% of respondents from sea grant reported focusing exclusively on tourism programming (>90% time commitment). A considerable percentage of respondents—47% of land grant respondents and 40% of sea grant respondents—reported allocating 10% or less of their time to tourism-related work (see Figure 2).

Despite the low percentage of time dedicated to tourism programs, respondents reported being engaged in a breadth of tourism-related program topical areas (Figure 3). While both land-grant and sea-grant Extension reported similar program offerings, there were notable differences in how they allocated time to these programs. For example, land-grant Extension professionals were more likely to engage traditional Extension audiences in rural destinations via agritourism initiatives (70% land grant vs 28% sea grant) and placemaking initiatives (21% vs 8%). In contrast, sea-grant activities were focused on coastal destinations and most likely to emphasize nature-based tourism (68% sea grant vs 27% land-grant); the environmental, social, and economic impacts of tourism activities (48% vs 25%); and education, training, and certification programs (44% vs 20%).

The tables in Appendix A show the combined land-grant and sea-grant topical areas related to tourism programs offered across the country from the states that responded to the survey.

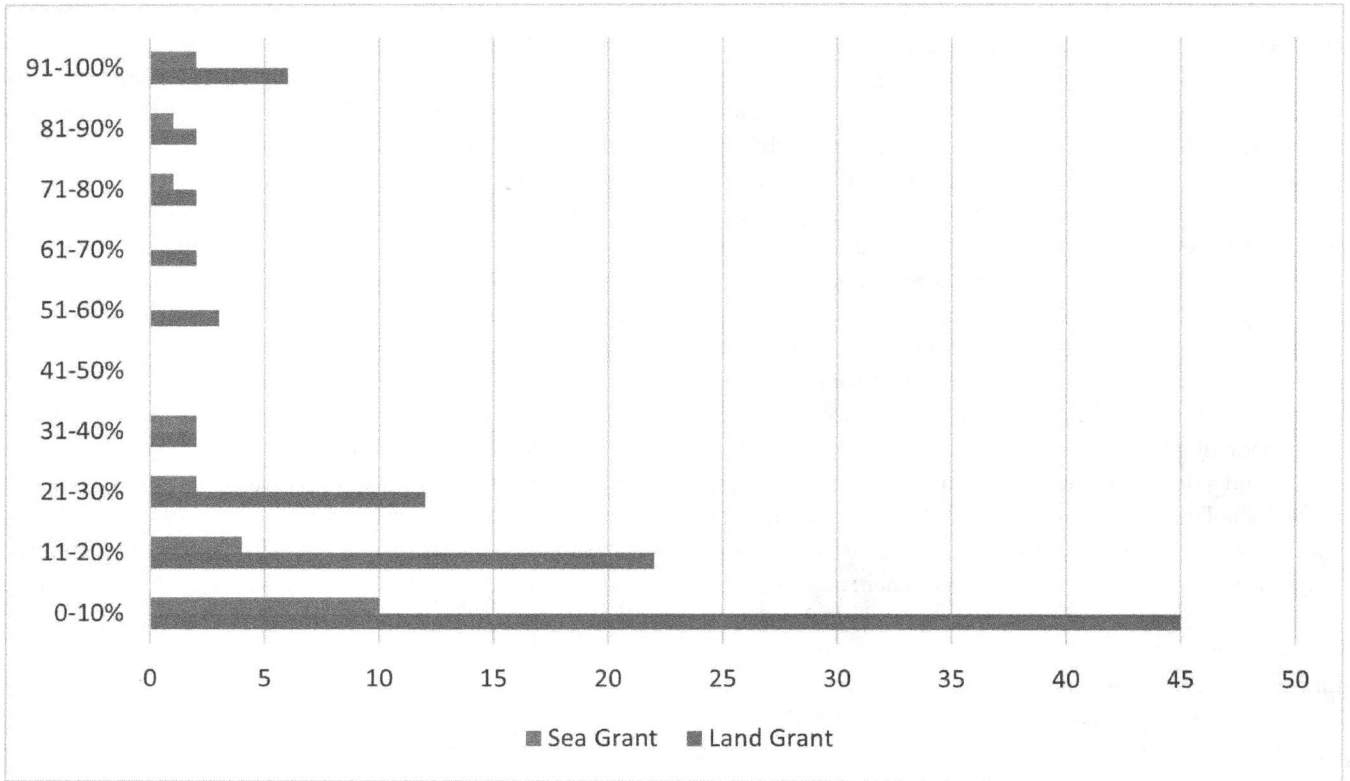

Figure 2. Percentage of time allocated to tourism-related work.

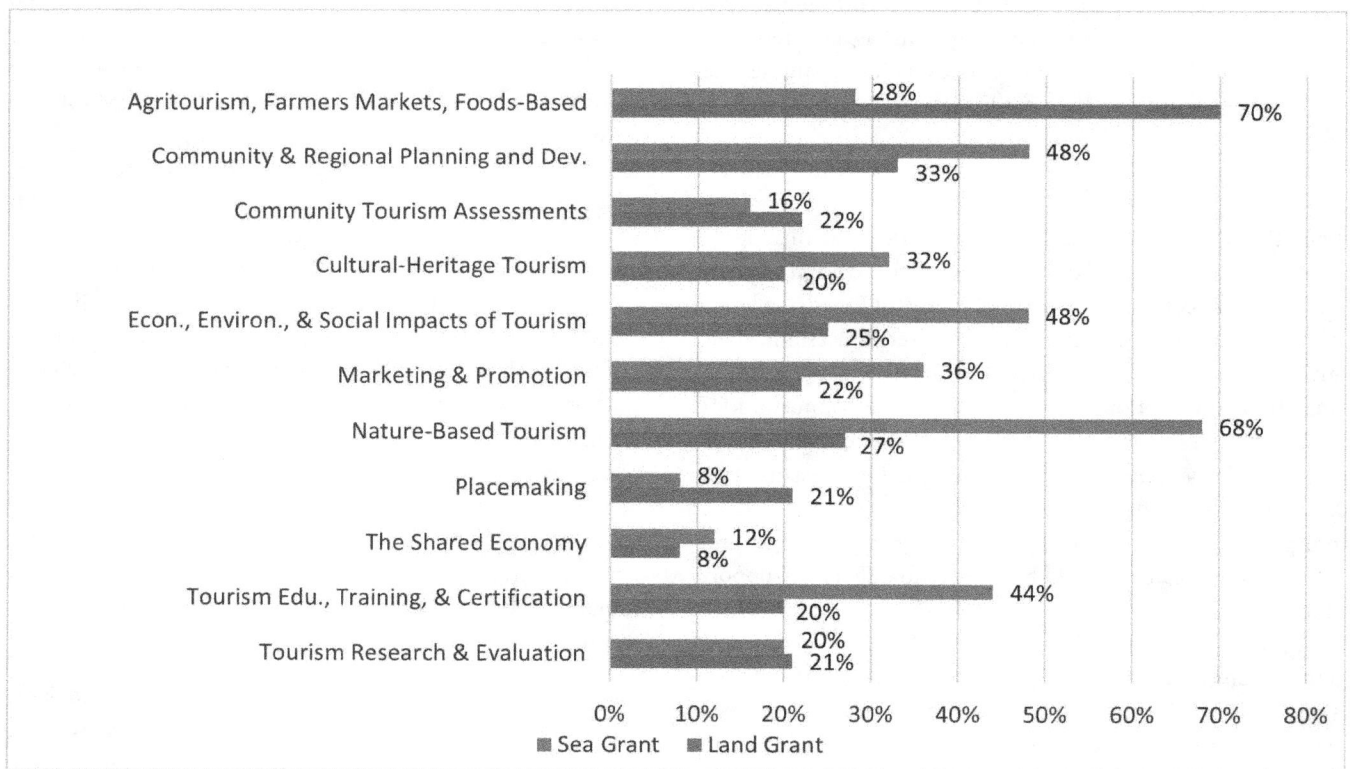

Figure 3. Extension tourism program topical areas offered or participated in. *Note.* Program descriptions and references to additional information are provided in the footnotes to Appendix A.

Although it is clear that Extension professionals are offering tourism programming, 87% of land-grant respondents and 77% of sea-grant respondents indicated that there is demand for tourism programming, but they are unable to meet the needs due to a lack of resources. Even among more traditional Extension educators for whom tourism is not their primary appointment, respondents noted a lack of human and monetary resources dedicated to tourism programming. For example, 24% of land-grant respondents cited a need for additional agritourism programs, suggesting high demand for this work. Land-grant Extension respondents also noted needs for programs related to nature-based eco-tourism; rural tourism; downtown business development; hospitality and customer service; mapping and community marketing of tourism resources; guide training; event management; economic impact analysis; recreation economies; visitor research; long-term planning; trips for youth, elderly, retirees, and handicapped; and tourism marketing programs.

Perhaps unsurprisingly, the programmatic needs identified by sea-grant professionals focused on coastal and maritime tourism needs such as the environmental impacts of cruise ships; marine environmental education for visitors; birding tourism; assessing the economic impacts of tourism on the coast; rural business development, especially for those businesses focused on ecotourism offerings and coastal field experiences; promoting and linking aquaculture and local food with tourism; conducting a needs assessment in the state; nature-based tourism and guide certification programs; courses related to sustainable tourism business practices and environmental issues/impacts; economic impact studies; discovery tours; leadership development for tourism professionals; and visitor carrying capacity studies.

Finally, respondents were asked to indicate the challenges or constraints they face in being able to provide services to support tourism in their state or region. For both land-grant and sea-grant Extension respondents, funding and staffing were identified as the greatest challenge or constraint with these challenges echoed equally among those with a high percentage of time allocated to tourism and those with a low percentage of time allocated to tourism-related programming. Although these are arguably challenges across Extension, it is notable that respondents also identified a lack of recognition and prioritization of their work. For example, respondents from land-grant institutions described tourism as not being a recognized Extension program and noted the lack of dedicated tourism programs and support at the university and state level. Respondents from both land-grant and sea-grant described cultural differences and communication challenges between tourism and traditional Extension community programs. Respondents also identified communication and coordination challenges between and across tourism development agencies and a lack of data to guide and direct programs as challenges.

CONCLUSION

Our results make it clear that tourism programming is an important component of the work conducted by Extension professionals across the country. This study demonstrates that although constrained by limited time, resources, and funding, Extension professionals are seeking ways to stay on the cutting edge of tourism development. However, despite its reach and continued and growing importance as an economic driver in rural communities, tourism continues to be underrepresented in CES programming. Indeed, the state of Extension tourism programming is largely unchanged over the past 25 years, and continues to be "substantial, although not at a level commensurate with the industry's economic importance." (The National Extension Travel and Tourism Advisory Committee, 1993).

Our study demonstrates the breadth and importance of Extension's tourism programming. We also highlight continued challenges, including limited investment and commitment by state institutions and the larger CES for core tourism program offerings. We continue to view investments in tourism programming as a way for Extension to maintain its relevancy and better engage and address the community and economic development needs of traditional and emerging audiences. The questions posed in 1967 by R.P. Davison, in 1978 by the CES Recreation and Tourism Task Force, and in 1993 by the National Extension Travel and Tourism Advisory Committee seem as increasingly relevant today as they were then: "Will the CES change its program priorities, organizational structure, and external relationships to meet the changing needs of society? What role will tourism play in the future of Extension?"

This study further identified that tourism provides a critical opportunity for Extension to engage new and existing constituents in relevant programming and warrants a change to CES priorities, investment, and organizational structures. Similar to previous efforts (The National Extension Travel and Tourism Advisory Committee, 1993), there are opportunities to establish tourism programming as a priority within CES, yet significant challenges to overcome to achieve this. The goal of these combined efforts remains the establishment of tourism as a recognized, supported program within the Cooperative Extension System and sea grant to meet emerging trends and the shifting needs of society, as illustrated in this assessment. The field's interdisciplinary nature and importance to local economies makes it not only increasingly relevant but also essential to Extension's constituents and worthy of increased investment by federal, state, and local partners. The National Extension Tourism network, established in 1993, is growing with new partners and strategic alliances in an attempt to embrace these opportunities and overcome the challenges identified in this study. Extension professionals and partners are invited to join the network at https://extensiontourism.net/.

REFERENCES

Boettner, F., Fedorko, E., Hansen. E., Collins, A., Zimmerman, B., Goetz, S. J., Han, Y., Gyovai, C., Carlson, E., & Sentilles, A. (2019) *Strengthening economic resilience in Appalachia. Two parts: Guidebook for practitioners and technical report.* Appalachian Regional Commission. https://www.arc.gov/wp-content/uploads/2019/02/StrengtheningEconomicResilienceGuidebook-Feb2019-1.pdf

Bureau of Labor Statistics (2019). *Quarterly Census of Employment and Wages.* [2019 Q3 Employment]. Retrieved from http://data.bls.gov

Chase, L. C., Stewart, M., Schilling, B., Smith, B., & Walk, M. (2018). Agritourism: Toward a conceptual framework for industry analysis. *Journal of Agriculture, Food Systems, and Community Development, 8*(1), 13–19. https://doi.org/10.5304/jafscd.2018.081.016

Destinations International (2019) *Destination NEXT futures study 2019: A strategic road map for the next generation of global destination organizations.* https://destinationsinternational.org/reports/destinationnext-futures-study

Future Tourism Development: Programming in the Cooperative Extension Service for the Next Millennium conference, Milwaukee, WI Sept. 1992.

Hargrove, C. (2020, May 13). *What is the difference between cultural tourism and heritage tourism?* Hargrove International. https://hargroveinternational.com/faqs/what-is-the-difference-between-cultural-tourism-and-heritage-tourism/

Highfill, T., Franks, C, & Georgi, P. S. (2018). Outdoor recreation satellite account: Updated Statistics. *Survey of Current Business 98*(9). https://apps.bea.gov/scb/2018/09-september/0918-outdoor-recreation.htm

Honadle, B. W. (1990). Extension and tourism development. *Journal of Extension, 28*(2). https://archives.joe.org/joe/1990summer/a1.php

Investopedia (2020, October 3). *Sharing economy.* https://www.investopedia.com/terms/s/sharing-economy.asp

Kusmin, L. (2016). *Using the ERS county economic types to explore demographic and economic trends in rural areas.* Amber Waves. https://www.ers.usda.gov/amber-waves/2016/december/using-the-ers-county-economic-types-to-explore-demographic-and-economic-trends-in-rural-areas/

Mission, vision, and strategic plan (n.d.). National Extension Tourism Network. https://extensiontourism.net/mission-vision-strategic-plan/

National Oceanic and Atmospheric Administration (2018, August 31). NOAA Sea Grant Coastal Tourism Vision Plan 2018-2028. https://seagrant.noaa.gov/Portals/1/FINAL%20%20Coastal%20Tourism%20Vision%20Plan%20.pdf

Project for Public Spaces (n.d.). *What is placemaking?* https://www.pps.org/category/placemaking

Rasker, R. (2018). *The economic value of federal public lands.* Headwaters Economics. https://headwaterseconomics.org/public-lands/public-lands-research/

Recreation and Tourism Task Force (1978). Recreation and Tourism: Challenges and Opportunities for Cooperative Extension. University of Minnesota Agricultural Extension Service. https://conservancy.umn.edu/handle/11299/169304

National Institute of Food and Agriculture (n.d.). *Regional rural development centers.* United States Department of Agriculture. https://nifa.usda.gov/regional-rural-development-centers

Sadowske, S. (1992). *Strategic initiatives for cooperative Extension in tourism and travel,* Executive Report of the National Extension Conference in Milwaukee, WI, September 1992.

Skift (2020). *Travel megatrends 2020.* https://skift.com/megatrends-2020/

Sjolander, S. A. (2020). *The virtuous cycle of placemaking.* Penn State Extension. https://extension.psu.edu/the-virtuous-cycle-of-placemaking

Texas Parks and Wildlife (n.d.). *What is nature tourism?* https://tpwd.texas.gov/landwater/land/programs/tourism/what_is/

The National Extension Travel and Tourism Advisory Committee (1993). *Tourism development: A suggested approach for the Cooperative Extension system.* https://extensiontourism.net/wp-content/uploads/2020/02/Tourism-Development-and-Cooperative-Extension-Report-1993.pdf

U.S. Bureau of Economic Analysis, Value Added by Industry as a Percentage of Gross Domestic Product. April 2020. Retrieved from https://apps.bea.gov/iTable/iTable.cfm?ReqID=51&step=1

Williams, M. R. & Landfried, T. (1992). *Communities in economic transition: Looking ahead to Extension program opportunities in community and economic development* [White Paper]. Cooperative Extension.

APPENDIX A. TOPICAL AREAS RELATED TO TOURISM PROGRAMS OFFERED BY STATE

Table A1.

Region	State	Nature-Based[a]	Agritourism[b]	Cultural-Heritage[c]	Shared Economy[d]	Placemaking[e]
North Central	IA	✓	✓	✓		
	IN	✓	✓	✓	✓	✓
	KS					
	MI	✓	✓	✓	✓	✓
	MN	✓	✓	✓	✓	✓
	ND		✓	✓		✓
	NE	✓	✓			
	OH	✓	✓	✓		✓
	WI					
Northeast	CT	✓	✓			✓
	DE					
	MA	✓				
	MD		✓			
	ME	✓	✓	✓		
	NH	✓	✓	✓	✓	✓
	NJ		✓			
	NY	✓	✓	✓		
	PA	✓	✓	✓	✓	✓
	RI		✓		✓	✓
	VT	✓	✓			
	WV	✓	✓	✓	✓	✓
South	AL	✓	✓	✓		✓
	AR		✓			
	FL	✓	✓	✓		
	GA	✓	✓			
	LA	✓				
	MS		✓	✓		✓
	NC	✓	✓	✓	✓	✓
	OK	✓	✓	✓		
	SC	✓	✓	✓		
	TN		✓			
	TX					
	VA		✓		✓	✓

Table A1. (*continued*)

Region	State	Nature-Based[a]	Agritourism[b]	Cultural-Heritage[c]	Shared Economy[d]	Placemaking[e]
West	AK	✓	✓			
	CA	✓	✓	✓		✓
	CO		✓			
	HI		✓			
	ID		✓	✓		
	MT	✓	✓			✓
	NM		✓		✓	
	NV	✓	✓			
	OR	✓	✓	✓	✓	
	WA	✓	✓	✓		✓

[a]Responsible travel to natural areas, which conserves the environment and improves the welfare of local people. (e.g., birdwatching, photography, stargazing, camping, hiking, hunting, fishing, and visiting parks) (Texas Parks and Wildlife, n.d.). [b]Visiting farms and ranches for education, hospitality, recreation, entertainment and/or direct sales of farm products (Chase et al., 2018). [c]"People-based" tourism that emphasizes engagement and learning of local traditions. (e.g., study tours, performing arts and cultural tours, travel to festivals, visits to sites and monuments, folklore, and pilgrimages) (Hargrove, 2020). [d]An economic model defined as a peer-to-peer based activity of acquiring, providing, or sharing access to goods and services that is often facilitated by a community-based online platform. (e.g., Airbnb, Uber, Zipcar, etc.) (Investopedia, 2020). [e]Placemaking is a participatory process that aims to improve a community's physical environment and the quality of human interactions (Sjolander, 2020). Placemaking facilitates creative patterns of use, paying particular attention to the physical, cultural, and social identities that define a place and support its ongoing evolution (Project for Public Spaces, n.d.).

Table A2.

Region	State	Education, Training, and Certification	Community Assessments	Planning and Development	Tourism Impacts	Marketing and Promotion	Research and Evaluation
North Central	IA	✓	✓			✓	✓
	IN	✓				✓	✓
	KS	✓					✓
	MI	✓	✓	✓			
	MN	✓	✓	✓	✓	✓	✓
	ND		✓	✓	✓	✓	
	NE	✓	✓	✓	✓	✓	✓
	OH	✓	✓	✓	✓	✓	✓
	WI			✓	✓		

Table A2. (*continued*)

Region	State	Education, Training, and Certification	Community Assessments	Planning and Development	Tourism Impacts	Marketing and Promotion	Research and Evaluation
Northeast	CT		✓	✓	✓		✓
	DE						
	MA			✓	✓		
	MD			✓			
	ME			✓		✓	
	NH	✓		✓	✓	✓	
	NJ					✓	✓
	NY	✓		✓	✓	✓	✓
	PA		✓	✓		✓	✓
	RI			✓	✓		
	VT				✓		✓
	WV	✓	✓	✓	✓		✓
South	AL		✓	✓	✓	✓	
	AR		✓				
	FL	✓	✓	✓	✓	✓	✓
	GA	✓			✓	✓	
	LA				✓	✓	
	MS	✓	✓			✓	✓
	NC	✓	✓	✓	✓	✓	✓
	OK	✓	✓	✓	✓		✓
	SC	✓		✓	✓	✓	✓
	TN						
	TX	✓					
	VA			✓		✓	✓
West	AK	✓			✓		
	CA	✓		✓	✓	✓	✓
	CO			✓	✓	✓	✓
	HI			✓	✓		✓
	ID		✓		✓		
	MT	✓		✓	✓	✓	✓
	NM	✓		✓			
	NV		✓	✓	✓	✓	
	OR	✓	✓	✓	✓	✓	✓
	WA	✓			✓	✓	✓

APPENDIX B. SURVEY RESPONSES BY STATE ACCORDING TO EXTENSION/RURAL REGIONAL DEVELOPMENT CENTER

Table B1. Survey Responses by State: Northeast Region

State	Land Grant	Sea Grant	Other	Total
CT	1	1	0	2
DE	0	1	0	1
DC	0	0	0	0
ME	2	1	0	3
MD	2	0	0	2
MA	1	0	1	2
NH	1	2	0	3
NJ	2	0	1	3
NY	1	1	0	2
PA	20	0	1	21
RI	0	1	0	1
VT	1	0	0	1
WV	5	0	0	5
TOTAL	**36**	**7**	**3**	**46**

Table B2. Survey Responses by State: Southern Region

State	Land Grant	Sea Grant	Other	Total
AL	12	1	1	14
AR	1	0	0	1
FL	0	4	0	4
GA	0	1	1	2
KY	0	0	0	0
LA	1	1	0	2
MS	1	0	0	1
NC	2	1	1	4
OK	2	0	0	2
SC	0	1	0	1
TN	1	0	0	1
TX	1	1	0	2
VA	2	1	1	4
TOTAL	**23**	**11**	**4**	**38**

Table B3. Survey Responses by State: Western Region

State	Land Grant	Sea Grant	Other	Total
AK	1	1	0	2
AZ	0	0	0	0
CA	2	0	0	2
CO	1	0	0	1
HI	1	1	0	2
ID	1	0	0	1
MT	1	0	3	4
NV	2	0	0	2
NM	2	0	0	2
OR	1	2	1	4
UT	0	0	0	0
WA	1	1	0	2
WY	0	0	0	0
TOTAL	**13**	**5**	**4**	**22**

Table B4. Survey Responses by State: Northcentral Region

State	Land Grant	Sea Grant	Other	Total
IL	0	0	0	0
IN	2	0	1	3
IA	2	0	0	2
KS	1	0	0	1
MI	3	1	0	4
MN	3	1	0	4
MO	0	0	0	0
NE	2	0	0	2
ND	1	0	0	1
OH	9	0	0	9
SD	0	0	0	0
WI	1	0	0	1
TOTAL	**24**	**2**	**1**	**27**

Journal of Extension

Feature Article

Volume 60, Issue 2, 2022

Overweight and Obesity Challenges Among African Americans in Rural Alabama Black Belt

JOEL TUMWEBAZE[1], JOHNPAUL KAGULIRE[1], AND NORMA L. DAWKINS[1]

AUTHORS: [1]Tuskegee University.

Abstract. This study aimed at averting challenges of overweight and obesity among African Americans in rural Alabama. Focus group discussions and surveys were used to design a 12-week nutrition education and physical activity program. Results showed a significant improvement ($p<0.05$) in participants' knowledge scores. Values for systolic blood pressure significantly reduced ($p<0.00$). There was an increase in number of participants with normal blood pressure and a reduction in number of participants with stage 2 hypertension. The study highlights that nutrition education and physical activity can lead to improved health outcomes among African Americans in rural Alabama.

INTRODUCTION

Obesity is characterized by excessive accumulation of fat in the body and is a risk factor for various chronic conditions, including diabetes, heart diseases, and certain types of cancer (World Health Organization, 2020). From 1999 to 2018, the prevalence of adult age-adjusted obesity in the United States increased by more than 10%, from 30.5% to 42.4% (Fryar et al., 2018; National Center for Chronic Disease Prevention and Health Promotion, 2020c). Notable disparities in obesity prevalence exist by race, ethnicity, education, age, and geographical distribution (Behavioral Risk Factor Surveillance System, 2020). In 2018, the obesity prevalence among African Americans was 49.6%, 44.8% among Hispanics, and 42.2% among Caucasians (Hales et al., 2020).

Geographically, southern states have the highest percentage of African Americans, and the average prevalence of obesity in the southern states was 33.6% in 2018 (Centers for Disease Control and Prevention [CDC], 2018). Alabama, a southern state whose population is approximately 25% African Americans, had the fifth highest (36.3%) adult obesity rate in 2018 and spends more than $5 billion dollars annually on obesity-related health conditions (Alabama Department of Public Health, 2020). Although there is limited literature on the prevalence of obesity in individual counties in Alabama, obesity rates in the Alabama Black Belt region exceed 40% for several counties, including Bullock County (National Center for Chronic Disease Prevention and Health Promotion, 2016; University of Wisconsin Population Health Institute, 2014).

According to the CDC, dietary lifestyle and physical activity continue to be the leading preventable risk factors for obesity (National Center for Chronic Disease Prevention and Health Promotion, 2020a). For instance, 32 % of adults in Alabama engaged in no physical activity during leisure time in 2020 compared to the national average of 26.6% (National Center for Chronic Disease Prevention and Health Promotion, 2020). Additionally, the percentage of adults in Alabama consuming fruits less than once per day in 2017 was 44.9%, a figure higher than the national average of 36% (National Center for Chronic Disease Prevention and Health Promotion, 2019). In Alabama, Bullock County has consistently exhibited poor health indicators such as adult obesity, low food environment index, physical inactivity, and socioeconomic factors such as poverty and poor education.

Several state and local programs have been instituted to improve physical activity and dietary lifestyles across the United States. These include the State Physical Activity and Nutrition (SPAN) Program, Racial and Ethnic Approaches to Community Health (REACH), and The High Obesity Program (Kahin et al., 2020; National Center for Chronic Disease Prevention and Health Promotion, 2020d). However, even with the interventions of such programs, the rates of overweight and obesity among African American communities remain high, indicating a need for more research.

Therefore, to determine and implement county-specific interventions needed to reduce overweight and improve nutrition among African Americans in Bullock County, Alabama, we (1) used healthy knowledge surveys and focus group discussions to identify barriers to healthy lifestyles and to identify content for nutrition education and physical activity modules and (2) determined the success of nutrition education and physical activity interventions using pre- and post-knowledge assessments and anthropometric data.

METHOD

RECRUITMENT OF PARTICIPANTS

A schematic summary of the study is represented in Figure 1 and expounded throughout the methods, results, and discussion sections that follow. The study received approval from the Institutional Review Board of Tuskegee University for use of human subjects. Participants were recruited from Union Springs, in Bullock County. Bullock County has among the highest percentage of obese adults, adults with diabetes, obese preschoolers, children in poverty, low-income families, and fast-food restaurants compared to other Alabama counties (University of Wisconsin Population Health Institute, 2021). Participants were recruited using flyers distributed to community centers, churches, and health centers by Tuskegee University Cooperative Extension agents through the snowballing technique. Snowball sampling is a method of recruiting participants where researchers do not

sample from a list of members of the population (known as a sampling frame). Instead, the recruited participant gives the researcher the name of at least one more potential participant, who in turn provides the name of at least one more potential participant, and so on, until the researcher obtains the required sample size (Bhattacherjee, 2012). Snowball sampling was used for our study because the rural population was hard to reach through other means of sample recruiting.

The criteria for eligibility included being African American, a resident of Bullock County, and between 18 and 65 years of age. Participants received an incentive of $10 for participating in the health knowledge surveys and focus group discussions and an incentive of $100 for participating in the nutrition education and physical activity program. The focus group discussions and health knowledge surveys were used to determine the content for the nutrition education and physical activity modules.

HEALTH KNOWLEDGE SURVEYS

One hundred ninety-two participants who consented to participate in the study were given health knowledge surveys, which they were requested to complete and drop at the Bullock County Extension office. We used a 33-item,

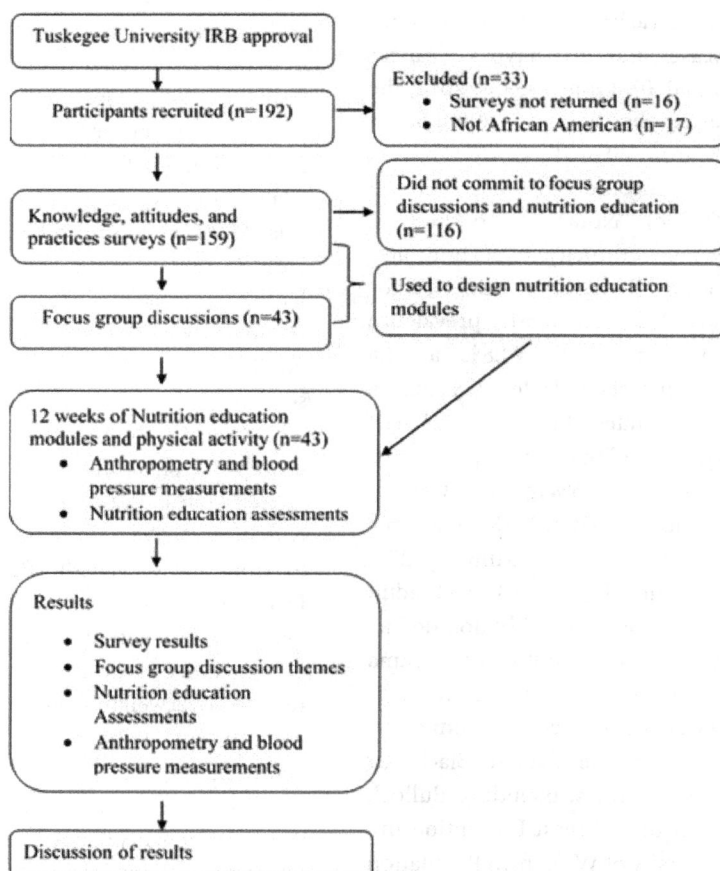

Figure 1. Schematic representation of the study.

self-administered survey adapted from Dawkins et al. (2010) that consisted of three sections: knowledge, attitudes, and perception. The knowledge section was comprised of closed-ended questions that assessed the participants' understanding of obesity, its prevalence, and its consequences. The perception and attitudes sections collected information on perceived strategies to fight obesity, perceived importance of obesity in relation to other diseases, and perceived roles of various stakeholders in preventing obesity.

A Likert-type four-point scale (strongly disagree, disagree, agree, and strongly agree) was used to determine participants' desire to learn more about obesity, various obesity prevention strategies, and the relationship between obesity and chronic diseases. This section also assessed participants' understandings of the roles played by the government, schools, employers, and healthcare providers in preventing obesity.

The survey was reviewed by Tuskegee University professors from the departments of psychology, nutrition, and education with knowledge in human behavior, human physiology, and human nutrition. After review, the survey was pretested for appropriateness in format, ethical considerations, logical order, and content using individuals from Bullock County. Subsequent tests for internal consistency reliability yielded a Cronbach's coefficient alpha of 0.85. Cronbach's coefficient measures how reliable the responses of a survey are, and it ranges from zero to one, with higher values indicating that surveys are measuring in the same dimension (Bujang et al., 2018). Participants who were given the surveys were requested to attend a focus group discussion one month after the survey was administered.

FOCUS GROUP DISCUSSIONS

Six focus group discussions consisting of a total of 43 participants were done to supplement the health knowledge surveys in determining barriers and opportunities for healthy lifestyles. The content of the focus group discussions was derived from participants' responses to the health knowledge surveys. Focus group discussions included 12 open-ended questions relating to participants' understandings of what constitutes a healthy lifestyle, the importance of physical activity, how social relationships relate to health, the role of community health programs, barriers and motivations to healthy lifestyles, and willingness to participate in nutrition education programs. The questions for the focus group discussions were pretested with five individuals from Bullock County who were not included in the study. After pretesting, the questions were formatted to ensure content comprehensiveness and ethical appropriateness.

All six focus group discussions were held in Bullock County, Alabama on separate weekdays in September 2019. Prior to the focus group discussions, the researchers explained the study to the participants and answered questions raised by the participants. Researchers then obtained written informed consent, and a copy of the signed consent form was given to each of the 43 participants. The six focus group discussions were delivered in English, moderated by a researcher, and lasted for 40 minutes. The discussions started with an icebreaker where each participant shared their name, favorite food, and for how long they had lived in Bullock County. The discussions were audio-recorded and supplemented with hand-written notes made by the research assistants.

NUTRITION EDUCATION MODULES AND PHYSICAL ACTIVITY

Using the barriers and opportunities for healthy lifestyles identified based on responses from the health knowledge surveys and focus group discussions, we developed content for the nutrition education and physical activity modules. Once a week for 12 weeks, researchers led participants through 60 minutes of nutrition education materials with topics including food nutrients, nutrient-disease interactions, food labeling, physical exercise, and strategies for behavior modification. Eight nutrition education modules were stretched out to cover the 12 weeks of the program.

After each day's nutrition education lesson, a licensed physical trainer led participants through 25 to 30 minutes of moderate intensity physical activities following guidelines from the American Heart Association (Piercy & Troiano, 2018; U.S. Department of Health and Human Services, 2018). In addition, participants were encouraged to continue physical activity at home for the 12-week duration of the program.

NUTRITION KNOWLEDGE AND ANTHROPOMETRY STATUS ASSESSMENTS

Nutrition Knowledge Assessments
At the beginning of each of the eight nutrition education and physical activity modules, participants were given a pre-knowledge assessment with 20 multiple-choice questions derived from content for that day. After the day's module, a post-knowledge assessment with the same questions as the pre-knowledge assessment was administered. The pre- and post-knowledge assessments were graded out of 100% and compared to determine the change in knowledge following the class.

Anthropometry and Blood Pressure Measures
Similarly, on the first and last days of the nutrition education and physical activity modules, anthropometry and blood pressure measurements were done and compared to determine changes in health status following the program. Procedures for measuring height, weight, hips, waist, and blood pressure were adapted from the Anthropometry Procedures Manual of the National Health and Nutrition Examination Survey (NHANES) (CDC, 2017). All measures were done by two research team members.

Height (cm) was measured using a portable stadiometer (Seca 213). Body weight (kg) was measured using a Detecto SlimPRO digital floor scale. Body mass index (BMI), an indirect measure of body fat, was calculated as a ratio of weight in kilograms to height in square meters. Waist and hip circumferences were measured in centimeters (cm) to the nearest 0.1 cm using a retractable measuring tape. The waist circumference was divided by the hip circumference to determine the waist to hip ratio (WHR).

Blood pressure was determined based on the method described by Muntner et al. (2019). For this method, we used an Omron blood pressure monitor (BP742N). Participants were required to sit up straight in a relaxed position while the researcher measured their blood pressure. Values of less than 120 mmHg for systolic and less than 80 mmHg for diastolic readings were used as a reference for normal blood pressure (Carey & Whelton, 2018; National Center for Chronic Disease Prevention and Health Promotion, 2020b).

STATISTICAL ANALYSIS

Quantitative Data Analysis

Researchers checked surveys for completion, coded them, and entered the responses into a Microsoft Excel database. Then, they imported data into Statistical Package for Social Sciences (SPSS). Descriptive statistics—including means, frequencies, percentages, and t-tests—were determined for the participants' survey responses. Researchers also analyzed data from nutrition education and anthropometry using Statistical Analysis Software (SAS). They then classified the participants' anthropometric and blood pressure measurements based on categorizations by the Centers for Disease Control and Prevention and the American Heart Association (National Center for Chronic Disease Prevention and Health Promotion, 2020a). Frequencies and percentages for anthropometry measures were determined, and the difference between means was compared using a paired t-test. All tests for quantitative data were done at a significance level of 0.05.

Qualitative Data Analysis

Audio recordings for focus group discussions were transcribed verbatim by the researchers. Using ATLAS.ti software (Version 8.4), researchers coded and analyzed the transcripts to reveal major themes arising from participants' responses.

RESULTS

PARTICIPANT RECRUITMENT

Out of the 192 surveys that were distributed to the recruited participants, 159 surveys were returned. Forty-three of the 159 participants who returned the surveys consented to participate in the focus group discussions, nutrition education, physical activities, and assessments.

HEALTH KNOWLEDGE SURVEYS

Results for the health knowledge surveys are presented in Tables 1 through 4. From the 192 total surveys given to the recruited participants, we excluded 33 surveys for participants who did not return the surveys and for participants who did not identify as African American.

Responses to general knowledge questions on obesity are presented as frequencies and percentages in Table 1. The results showed that a majority of the participants understood that the prevalence of obesity was high and that it was caused by lifestyle and genetic factors. Additionally, the data revealed that a majority of participants had some knowledge on ways to prevent or reduce obesity. Most participants (89.3%) regarded obesity as reaching epidemic levels, viewed obesity as the result of an abundance of caloric intake (83%), and observed that there were multiple risk factors for obesity including genetic, socioeconomic, environmental, and behavioral factors (84.3%).

Similarly, survey results showed that a majority of participants were knowledgeable about obesity being a serious public health problem (90.6%) and a risk factor for chronic health conditions (Table 2). Furthermore, 88.1% agreed that the prevalence of obesity among children was increasing, as was their vulnerability to type 2 diabetes (80.5%).

We cross tabulated participants' views on the different obesity prevention and reduction programs with their education level (Table 3). This was done to determine whether education influences participants' views on obesity and whether nutrition education was necessary. Participants were asked whether they supported or opposed various efforts (table 5) to fight obesity in children. Results showed that education level had an impact on participants' views on interventions to reduce obesity. There was a significant difference in the views of college and high school graduate participants pertaining to the prohibition of selling unhealthy foods in vending machines ($p < 0.00$), limiting unhealthy food ads to children ($p < 0.02$), and providing more physical activity in schools ($p < 0.00$).

Similarly, the annual incomes of participants were cross tabulated against participants' views on obesity reduction programs. Generally, annual income did not affect participants' views of obesity reduction programs (Table 4). However, support for "limit[ing] TV ads for unhealthy foods and drinks" was more significantly ($p > 0.04$) supported by participants who earned more than $40,000 compared to those who earned less.

Survey results show that education may be used to increase participants' knowledge of obesity reduction strategies. Increasing participants' incomes may not be a suitable intervention for improving participants' perceptions of obesity and obesity reduction strategies.

Overweight and Obesity Challenges Among African Americans in Rural Alabama Black Belt

Table 1. Participants' Responses to General Knowledge Statements on Overweight and Obesity

Statement	True n (%)	False n (%)	No opinion n (%)
Obesity and overweight are reaching epidemic levels.	142 (89.3)	16 (10.1)	1 (0.6)
Obesity and overweight affect approximately 72% of US adults.	132 (83.0)	22 (13.8)	5 (3.2)
Obesity and overweight are primarily due to people consuming more calories than they burn through physical activity.	141 (88.7)	18 (11.3)	–
The problem of obesity is a combination of genetic, metabolic, behavioral, environmental, and socioeconomic factors.	134 (84.3)	22 (13.8)	3 (1.9)
Adding a moderate amount of exercise five or more times a week can lead to substantial weight loss.	131 (82.4)	28 (17.6)	–
Low carbohydrate diets and weight loss pills are the fastest ways to lose weight.	34 (21.4)	120 (75.5)	5 (3.1)
Obesity and overweight are the second leading causes of preventable death in the United States, close behind tobacco use.	109 (68.6)	46 (28.9)	4 (2.5)
Setting nutritional goals alone will help one lose weight.	91 (57.2)	66 (41.5)	2 (1.3)
"Understanding what and why you eat" is a major factor for implementing weight control.	150 (94.3)	7 (4.4)	2 (1.3)
Eating a snack before bedtime is a good way to implement weight control.	19 (11.9)	137 (86.2)	3 (1.9)
At age 21, males and females stop growing and no longer need to implement weight control.	14 (8.8)	142 (89.3)	3 (1.9)
Knowing your BMI is important in predicting whether you are overweight or obese.	135 (84.9)	22 (13.8)	2 (1.3)
It is easier to lose weight when overweight than when morbidly obese.	55 (34.6)	94 (59.1)	10 (6.3)
Overeating and inadequate physical activity are the main causes of obesity.	131 (82.4)	20 (12.6)	8 (5.0)

Note. n = 159. *Numbers in parenthesis indicate percentages of participants for each category.

Table 2. Knowledge on Association of Obesity and Chronic Disease

Statement	Participant responses, n (%)		
	Yes	No	No opinion
Overweight and obesity are risk factors for other diseases.	144 (90.6)	11 (6.9)	4 (2.5)
There is an increased rate of type 2 diabetes among children and teenagers.	128 (80.5)	25 (15.7)	6 (3.8)
The rate of overweight and obesity is greater in children than in adults.	76 (47.8)	73 (45.9)	10 (6.3)
Overweight and obesity are a potential health problem.	144 (90.6)	12 (7.5)	3 (1.9)
The CDC estimates that about 6 out of 10 Americans are overweight and obese.	115 (72.3)	37 (23.3)	7 (4.4)
Overweight and obesity in children is higher than in previous years.	140 (88.1)	14 (8.8)	5 (3.1)

Note. n = 159. *Numbers in parenthesis indicate percentages for each category.

Table 3. Effect of Education Level on Participants' Support for Obesity Reduction

Statement	High School	College	t-test	p-value
Prohibit the sale of soda, chips, and candy in school vending machines.	2.58±1.14	3.19±0.88	-3.07	0.00*
Limit TV ads for unhealthy foods and drinks (soda, chips, candy) that are targeted at children.	2.77±1.08	3.21±0.86	-2.28	0.02*
Educate parents about childhood obesity and healthier eating and exercise habits for children.	3.69±0.47	3.85±0.46	-1.53	0.13
More physical activities in schools.	3.56±0.51	3.89±0.39	-3.65	0.00*
Provide healthier school lunches.	3.48±0.71	3.69±0.58	-1.56	0.12

Note. n = 159. Responses were ranked in a Likert scale where 4 = *strongly support*, 3 = *somewhat support*, 2 = *somewhat oppose*, and 1 = *strongly oppose*. Statistically significant differences exist at p ≤ 0.05.

Table 4. Effect of Income on Participants' Support for Obesity Reduction

Statement	< $40,000	> $40,000	t-test	p-value
Prohibit the sale of soda, chips, and candy in school vending machines.	3.08±0.94	3.22±0.94	-0.87	0.39
Limit TV ads for unhealthy foods and drinks (soda, chips, candy) that are targeted at children.	3.02±0.98	3.36±0.78	-2.11	0.04*
Educate parents about childhood obesity and healthier eating and exercise habits for children.	3.79±0.44	3.86±0.47	-0.82	0.42
More physical activities in schools.	3.86±0.35	3.86±0.47	-0.07	0.95
Provide healthier school lunches.	3.66±0.54	3.75±0.54	-0.91	0.36

Note. n = 159. Responses were ranked in a Likert scale where 4 = *strongly support*, 3 = *somewhat support*, 2 = *somewhat oppose*, and 1 = *strongly oppose*. Statistically significant differences exist at p ≤ 0.05.

FOCUS GROUP DISCUSSIONS

Participants who took part in the surveys were invited to be part of focus group discussions and the nutrition education course. Of the 159 participants who completed the health knowledge surveys, 43 participants consented to take part in focus group discussions, and they were divided into six groups. Through the discussions, six themes were identified which represented barriers and opportunities for designing health interventions. The six themes (barriers and opportunities) are summarized in Table 5 and discussed below.

Physical Activity

Participants mentioned that regular physical exercise regimens such as walking and swimming were necessary for managing body weight, strengthening the immune system, reducing emotional stress, strengthening muscles, minimizing pain in body joints, attaining good sleep, and maintaining physical fitness, youthfulness, and vitality. However, participants reported that maintaining a physically active lifestyle in Bullock County was hampered by limited environments for physical activities, including walking and biking trails, parks, and gyms. Additionally, in areas where walking trails and parks were available, participants felt unsafe to exercise without a group. On an individual basis, barriers to regular physical activity included physical disabilities, tight work schedules, lack of commitment, limited social support, and lack of motivation.

Dietary Intake

Participants expressed that establishing good dietary habits involves minimizing the consumption of refined foods. Some explained that working multiple jobs for long hours made planning and making homemade meals difficult. Instead, participants reported depending on convenience foods that require little or no preparation, including already prepared fast foods, canned foods, and sugar-sweetened beverages.

Food Access

Limited access to fresh produce was cited as a barrier to making healthy choices. This was partly due to the absence of large, chain grocery stores like Walmart, Publix, and Costco in their communities. As such, participants reported that residents in need of fresh groceries drive long distances to neighboring cities—a costly practice in terms of transportation and time. As a result, residents relied on convenience stores within their communities shelved with high sodium canned foods, high calorie refined snacks, and a limited selection of fruits.

Social Influence

Participants reported that social networks influenced their dietary and lifestyle choices. It was mentioned that the cohesiveness of a family played an important role in developing and maintaining healthy living habits, especially for the children. Additionally, they emphasized that depression promotes unhealthy eating habits, but friends and family provide the motivation and social support necessary for managing or preventing depression.

Attitudes, Cultures, and Beliefs

Participants expressed that they believed their health was determined by the quality of their diet. Their responses revealed that healthy mindsets were greatly influenced by social relations, especially relationships with their family and friends. Participants explained that most of the foods and methods of cooking they used were passed down from their parents and grandparents. They mentioned that African American cuisine grew out of the resourcefulness of their ancestors during slavery as they learned to combine their traditional cuisines with that of their masters. One respondent explained how their ancestors were creative to use parts of the food that would otherwise be discarded such as the snout, stomach, and small intestines of pigs, which,

Table 5. Quotes From Focus Group Discussions in Bullock County, 2019

Theme	Sample quotes from focus group discussions
Physical activity	"When you start routine exercise, it really helps. I exercise twice a day: mornings and evenings. I believe that staying on a routine exercise regime will help me live longer." "Working out is hard at our age, but I was told that I can exercise in my couch while seated, and this can be good for the strength of my upper body and mind." "I need to have a stronger memory."
Dietary habits	"Changing dietary habits is a concern to me today since there is a lot of consumption of fast foods." "Somebody from a community program led by Tuskegee University came here and cautioned us on how to eat right. She said, 'Take that soda off your table.' And since then, I have kept it off my table."
Food access	"There is limited access to large grocery stores like Walmart, Publix, and Costco, making it difficult to get fresh and quality fruits and vegetables."
Social relations	"Hanging out with friends affects our lives—for the good or bad. For example, hanging out around smokers may lead to smoking." "If you have a lazy partner, you may be discouraged from carrying out any form of exercise."
Attitudes, perceptions, and beliefs	"Some people bring negativity, and so it is best to cut them out of your circle even if they are family." "It is highly perceived that eating fresh and healthy foods is expensive, especially for people in my community."
Healthcare and healthy lifestyle interventions	"I regularly have medical checkups to monitor my blood pressure, cholesterol, and blood sugar levels." "Our representatives should ask for the establishment of fresh produce grocery stores, improving the support environment for physical activities, and continued support through nutrition education and physical activity extension services." "We need to do routine medical screening for blood pressure and diabetes, but insurance is another issue here in our community."

when boiled, are referred to as chitterlings—a commonly used food in rural African American communities.

Healthcare and Healthy Lifestyle Interventions

Several of the participants reported that they were conversant with the harmful effects of overconsumption of certain nutrients as well as the benefits of eating in moderation. Participants reported that they had regular medical checkups and that they monitored their blood pressure, cholesterol, and blood sugar levels. However, some participants reported that they needed to have an expensive health insurance to keep up with medical checkups. Participants suggested interventions that would lead to attaining and maintaining healthy lifestyles. These interventions included asking their government leaders to lobby for the establishment of fresh produce grocery stores, improving the support environment for physical activities, and continuing the availability of nutrition education and physical activity extension services. Participants reported that the community should have more health and wellness services for the retired elderly and more physical activity programs for school children. In addition, participants suggested the need for engaging more local stakeholders, such as schools, churches, and community centers, in health awareness services.

NUTRITION EDUCATION AND ASSESSMENTS

The barriers and opportunities identified through the health knowledge surveys were supplemented with themes from the focus group discussions to design content for eight nutrition education modules. All participants that took part in the focus group discussions (n = 43) participated in the 12 weeks of nutrition education and the corresponding nutrition education and physical activity assessments. The percentage change in scores from assessments done before and after each interactive module represented change in nutrition knowledge. Results showed significant changes in nutrition knowledge for all modules except the food labeling module (p = 0.32) (Table 6). The percentage change in knowledge nearly doubled for the vitamins module (45.8%) but was lowest for the food labeling module (6.6%).

ANTHROPOMETRY AND BLOOD PRESSURE MEASUREMENTS

Anthropometry and blood pressure measurements were done at the beginning of the first module and at the end of the last module. These were done to determine the effectiveness of the nutrition education and physical activities. Results showed a decrease in average weight, waist circumference, and hip circumference at post-study. However, there were no significant differences (p > 0.05). The number of participants in the normal, overweight, and obese BMI categories remained unchanged from baseline to post-study. The average systolic pressure of participants significantly reduced (p < 0.00) from 135.7±16.7 mmHg at baseline to 123.1±26.0 mmHg at post study (Table 7). However, there was no significant difference in participants' diastolic pressure (p < 0.13).

DISCUSSION

FOCUS GROUP DISCUSSIONS AND HEALTH KNOWLEDGE SURVEYS

It is not clear why only 27% of the participants that took part in the health knowledge surveys consented to focus group discussions, nutrition education, physical activities, and the corresponding assessments. From both the focus group discussions and the health knowledge surveys, most participants recognized obesity as an epidemic with negative health consequences and observed that obesity could be averted through diet management practices and

Table 6. Pre- and Post-Test Scores for Nutrition Education Module Assessments

Nutrition module	Pre-test (%)	Post-test (%)	% Change	p-value
Carbohydrates	67.4±15.4	75.9±19.6	12.6	0.03
Proteins	59.1±17.5	73.4±17.9	24.2	0.00
Lipids	70.0±21.7	92.1±9.8	31.6	0.00
Fruits and vegetables	68.5±9.4	81.7±9.7	19.3	0.00
Vitamins	58.5±19.7	85.3±14.8	45.8	0.00
Minerals	59.3±19.6	76.3±20.0	28.7	0.00
Food labeling	80.7±34.6	86.0±22.4	6.6	0.32
Physical activity	54.1±20.1	74.4±18.5	37.5	0.00

Note. Values represent Mean±SD (n = 50). Statistical significance considered at p ≤ 0.05.

Table 7. Anthropometry and Blood Pressure Characteristics of Participants

Characteristic	Baseline	Post-study	p-value
Weight (kg)	95.3±21.1	94.2±20.7	0.39
Waist circumference (cm)	109.7±15.2	108.3±13.8	0.09
Hip circumference (cm)	121.3±16.5	120.3±16.7	0.16
Waist Hip Ratio (WHR)	0.905±0.07	0.901±0.06	0.69
Body Mass Index (BMI)			
BMI (kg/m²)	36.1±8.5	36.4±8.5	0.28
Blood Pressure (BP) mmHg			
Systolic BP (mmHg)	135.7±16.7	123.1±26.0	0.00
Diastolic BP (mmHg)	78.8±10.3	84.2±17.6	0.13

regular physical activities. Diet management and physical activity have been identified to promote individual health and wellbeing in the Black Belt region by other researchers (Carter et al., 2015; Scarinci et al., 2014). The knowledge on obesity displayed by the participants is important because, being African Americans, all our participants are genetically predisposed to obesity and its underlying chronic conditions, including certain types of cancer, diabetes, and cardiovascular diseases (Kumanyika et al., 2014).

The high levels of knowledge about obesity observed among participants may be due to higher prevalence of the condition in the Alabama Black Belt region and previous efforts to increase obesity awareness among these communities (Carter et al., 2010; Carter et al., 2015; CDC, 2018). Studies have shown that the possibility of ill health due to advanced age and having ample knowledge may be motivational factors for behavior change. As such, the advanced age (63.5±8.9) of participants in the present study and the high knowledge of obesity they displayed may have been motivational factors for participation. This logic is supported by the health belief model, which highlights that a person's perceived risk of developing a health condition and its consequences may be motivational factors for behavioral change (Janz & Becker, 1984).

Responses further revealed that participants' food choices were influenced by dietary habits passed down through generations of African American descendants. Participants reported the consumption of soul food—a traditional African American cuisine commonly characterized by the centrality of pork and developed due to racial stigma, resourcefulness, ingenuity, and communal spirit (Miller, 2013; Smith et al., 2006). Other researchers have identified pork-based meals, salted meats, and adding butter during vegetable preparation as common components of cooking an African American diet (Bovell-Benjamin et al., 2009). Due to the specific diet identified in our study, strategies used to promote healthy lifestyles in other geographical areas may not be applicable in the Alabama Black Belt counties. Other researchers have recommended strategies to implement dietary practices that are geographically and culturally specific (Bovell-Benjamin et al., 2009; Scarinci et al., 2014).

The barriers and recommendations identified through the focus group discussions and health knowledge surveys in this study led to the development of nutrition education modules tailored to the participants. Among other topics, the developed modules focused on cooking methods, the importance of fruits and vegetables in the diet, and the harmful effects of high sodium and saturated fat in the diet.

NUTRITION EDUCATION ASSESSMENT CONSIDERATIONS

The average pre-test scores for all nutrition education assessments were over 50%, indicating that participants had an average understanding of the topics presented, including food nutrients, food choices, physical activity, and food preparation. However, on average, pre-test scores for food labeling were higher than those from other modules.

The higher scores for food labeling compared to other modules at baseline may be attributed to participants' interests in sodium and calories, as they relate to blood pressure and obesity, respectively. During focus group discussions, participants were concerned about the intake of excess sodium and calories from refined foods and convenience foods readily available at grocery stores in their community. On most nutrition facts labels, especially labels of canned and convenient foods, information about calories and sodium is more pronounced. This may be the reason why participants were more knowledgeable about food labels.

Even though changes in scores for carbohydrates, proteins, minerals, and physical activity were significant, the average post-score value was 70%. Participants performed worst on questions that required quantitative skills, including calorie counting, understanding daily values, and interpreting serving size. For example, most of the incorrect responses on the assessments for the food labeling, fruits and vegetables, and proteins modules required quantitative responses. There was a greater change in knowledge scores for modules that required qualitative responses—such as the functions of different vitamins, types of physical exercises, and types and sources of nutrients—compared to quantitative responses. Other researchers reported that participants had difficulty comprehending quantitative information displayed on food labels, including serving sizes and recommended daily values (Campos et al., 2011). Researchers explained that most consumers could easily locate caloric content information on nutrition labels (Sinclair et al., 2013), but fewer consumers reported using percentage daily values and serving sizes to estimate their required daily nutrient intake (Levy & Fein, 1998; Moore et al., 2018).

Understanding numerical values on food labels may be a key factor for ensuring recommended nutrient intakes. For our study, understanding quantitative information was crucial given that during focus group discussions, participants in our study reported consumption of fast foods and convenient foods. However, it was important that we obtained significant improvement in nutrition knowledge scores. Educational interventions have led to improvement in nutrition knowledge; in 2018, Moore et al. reported that all 17 studies included in a systematic review showed a statistically significant improvement in participants' understanding of nutrition labels.

ANTHROPOMETRY CONSIDERATIONS

Participants' overall reduction in weight during this study, although non-significant, may signal that a longer intervention duration could lead to more weight reductions and possibly a decline in BMI. Carter et al. (2015) reported significant reductions in weight and BMI after delivering nutrition education and physical activity interventions in a similar population for a longer period with multiple focused physical activities. The rate of obesity reported in our study is consistent with that reported by the Centers for Disease Control and Prevention (CDC, 2018; Tumwebaze et al., 2021). Although this similarity in results demonstrates reproducibility, it unfortunately shows that the rate of obesity remained unchanged between 2018 and 2019. On a positive side, the unchanged rate of weight gain may signal an end to the rising trend in obesity that has been consistent for this population in the last two decades.

However, some researchers found African Americans (specifically, women) to be more self-accepting of weight, body shape, and appearance than white women (Abrams et al., 1993; Akan & Grilo, 1995; Befort et al., 2008; Parham-Payne, 2013). As such, African American women may have beliefs that perpetuate a cultural preference for heavier figures. This may be a reason for the non-significant changes

in BMI for our study. Therefore, strategies that do not focus on weight loss but rather on healthy eating and physical activity may be a more promising approach for promoting a healthy lifestyle (Befort et al., 2008; Willet & Stampfer, 2013; Woll et al., 2013).

BLOOD PRESSURE

Race is among the risk factors for high blood pressure reported by several researchers (Carey & Whelton, 2018; Willett & Stampfer, 2013). Because participants in our study were African Americans, a race associated with a risk for high blood pressure, this may partially explain why the blood pressure values from our study were elevated.

A significant decline in systolic blood pressure, a decline in number of participants with stage 2 hypertension, and an increase in number of participants with normal blood pressure was seen during this study. Results from our study showed that the mean blood pressure value was within the elevated blood pressure category, an interesting finding given that over 25% of African Americans were reported by other researchers to have high blood pressure (National Center for Chronic Disease Prevention and Health Promotion, 2010). Topics from our nutrition education modules such as monitoring sodium, cholesterol, and water intake along with having regular medical checkups could have directly influenced the observed blood pressure results.

LIMITATIONS

The sample size for focus group discussions, nutrition education, and physical activities in this study is generally considered small for statistical purposes. Similarly, the snowballing method of sampling may not present a representative sample. Based on population trends, Bullock County has seen negative population growth over recent years, with fewer individuals available and willing to participate in community development studies. Thus, snowball sampling was required to recruit available participants. Our study involved qualitative research, so it is possible that during focus group interviews, participants could have aligned their responses with those that are socially acceptable rather than what they thought was the truth. However, we minimized the possibility of social desirability bias at the beginning of the focus group interviews and nutrition education modules by establishing rapport with the participants and by following their responses with clarifying questions.

CONCLUSION

The present study used information from focus group discussions and healthy knowledge surveys to design nutrition education modules and physical activities for residents in Bullock County, Alabama. The significant change in knowledge scores is important because African Americans are more genetically predisposed to obesity-related chronic diseases; strategies aimed at increasing knowledge and changes in lifestyle are likely to reduce the risk for chronic diseases. Low scores from the nutrition labeling assessments may present an opportunity for researchers to develop a simplified way to teach how to understand the food labels or to develop a simplified food labels.

The high levels of obesity observed by the present study and by similar studies in the Alabama Black Belt region highlight a need to reevaluate current obesity prevention interventions. To develop effective overweight and obesity reduction strategies, researchers and policy makers may need to understand the knowledge, attitudes, practices, barriers, and opportunities that are specific to members in the targeted community. It is thought that a longer study duration would result in significant changes in weight and, thus, in BMI. Long term weight reduction and maintenance may require multiple continued nutritional, educational, and environmental interventions. Although the United States has several state and local programs to tackle obesity, our study highlights the need for culturally and geographically specific interventions.

REFERENCES

Abrams, K. K., Allen, L. R., & Gray, J. J. (1993). Disordered eating attitudes and behaviors, psychological adjustment, and ethnic identity: A comparison of black and white female college students. *International Journal of Eating Disorders, 14*(1), 49–57.

Akan, G. E., & Grilo, C. M. (1995). Sociocultural influences on eating attitudes and behaviors, body image, and psychological functioning: A comparison of African American, Asian-American, and Caucasian college women. *International Journal of Eating Disorders, 18*(2), 181–187.

Alabama Department of Public Health. (2020). *Obesity trends (data)*. Bureau of Prevention, Promotion, and Support. https://www.alabamapublichealth.gov/obesity/trends.html

Befort, C. A., Thomas, J. L., Daley, C. M., Rhode, P. C., & Ahluwalia, J. S. (2008). Perceptions and beliefs about body size, weight, and weight loss among obese African American women: A qualitative inquiry. *Health Education & Behavior, 35*(3), 410–426.

Behavioral Risk Factor Surveillance System. (2020. *Prevalence of self-reported obesity among U.S. adults by state and territory*. Centers for Disease Control and Prevention. https://www.cdc.gov/obesity/data/prevalence-maps.html

Bhattacherjee, A. (2012). *Social science research: Principles, methods, and practices*. Textbooks collection. Book 3, p.2015.

Bovell-Benjamin, A. C., Dawkin, N., Pace, R. D., & Shikany, J. M. (2009). Use of focus groups to understand African-Americans' dietary practices: Implications for modifying a food frequency questionnaire. *Preventive Medicine, 48*(6), 549–554.

Bujang, M. A., Omar, E. D., & Baharum, N. A. (2018). A review on sample size determination for Cronbach's alpha test: A simple guide for researchers. *The Malaysian Journal of Medical Sciences, 25*(6), 85.

Campos, S., Doxey, J., & Hammond, D. (2011). Nutrition labels on pre-packaged foods: A systematic review. *Public Health Nutrition, 14*(8), 1496–1506.

Carey, R. M., & Whelton, P. K. (2018). Prevention, detection, evaluation, and management of high blood pressure in adults: Synopsis of the 2017 American College of Cardiology/American Heart Association Hypertension Guideline. *Annals of Internal Medicine, 168*(5), 351–358.

Carter, V. L., Dawkins, N. L., & Howard, B. (2010). Healthy lifestyle: A community-based cancer awareness and prevention intervention program. *Journal of Health Care for the Poor and Underserved, 21*(3), 107–118.

Carter, V. L., Dawkins, N. L., & Howard, B. (2015). Weight and blood pressure reduction among participants engaged in a cancer awareness and prevention program. *Preventive Medicine Reports, 2*, 858–861.

Centers for Disease Control and Prevention. (2018). *NHANES 2017–2018 procedure manuals*. National Center for Health Statistics. https://wwwn.cdc.gov/nchs/nhanes/continuousnhanes/manuals.aspx?BeginYear=2017

Dawkins, N., McMickens, T., Findlay, H., & Pace, R. (2010). Community leaders' knowledge and perceptions about obesity: Implications for outreach educators in designing interventions. *Journal of Extension, 48*(5), 1–13. https://tigerprints.clemson.edu/joe/vol48/iss5/17/

Fryar, C. D., Carroll, M. D., & Ogden, C. (2018). *Prevalence of overweight, obesity, and severe obesity among adults age 20 and over: United States, 1960–1962 through 2015–2016*. National Center for Health Statistics. https://www.cdc.gov/nchs/data/hestat/obesity-adult-17-18/obesity-adult.htm

Hales, C.M., Carroll, M.D., Fryar, C.D., & Ogden, C.L. (2020). Prevalence of obesity and severe obesity among adults: United States, 2017-2018. NCHS data brief, number 360.

Janz, N. K., & Becker, M. H. (1984). The health belief model: A decade later. *Health Education Quarterly, 11*(1).

Kahin, S. A., Murriel, A. L., Pejavara, A., O'Toole, T., & Petersen, R. (2020). The High Obesity Program: A collaboration between public health and cooperative extension services to address obesity. *Preventing Chronic Disease, 17:E26*

Kumanyika, S. K., Whitt-Glover, M. C., & Haire-Joshu, D. (2014). What works for obesity prevention and treatment in black Americans? Research directions. *Obesity Reviews, 15*, 204–212.

Levy, A. S., & Fein, S. B. (1998). Consumers' ability to perform tasks using nutrition labels. *Journal of Nutrition Education, 30*(4), 210–217.

Miller, A. (2013). *Soul food: The surprising story of an American cuisine, one plate at a time.* UNC Press.

Moore, S. G., Donnelly, J. K., Jones, S., & Cade, J. E. (2018). Effect of educational interventions on understanding and use of nutrition labels: A systematic review. *Nutrients, 10*(10), 1432.

Muntner, P., Shimbo, D., Carey, R. M., Charleston, J. B., Gaillard, T., Misra, S., & Wright Jr., J. T. (2019). Measurement of blood pressure in humans: A scientific statement from the American Heart Association. *Hypertension, 73*(5), e35–e66.

National Center for Chronic Disease Prevention and Health Promotion. (2010). *A closer look at African American men and high blood pressure control: A review of psychosocial factors and systems-level interventions.* Centers for Disease Control and Prevention. https://www.cdc.gov/bloodpressure/docs/african_american_sourcebook.pdf

National Center for Chronic Disease Prevention and Health Promotion. (2016). *Programs to reduce obesity in high obesity areas.* Centers for Disease Control and Prevention. https://www.cdc.gov/nccdphp/dnpao/state-local-programs/profiles/15_262070a_reduceobesity_al_fs_final2_508tagged.pdf

National Center for Chronic Disease Prevention and Health Promotion. (2019). *Data, trends, and maps.* Centers for Disease Control and Prevention. https://www.cdc.gov/nccdphp/dnpao/data-trends-maps/index.html

National Center for Chronic Disease Prevention and Health Promotion. (2020a). *Adult obesity causes and consequences.* Centers for Disease Control and Prevention. https://www.cdc.gov/obesity/adult/causes.html

National Center for Chronic Disease Prevention and Health Promotion. (2020b). *Facts about hypertension.* Centers for Disease Control and Prevention. https://www.cdc.gov/bloodpressure/facts.html

National Center for Chronic Disease Prevention and Health Promotion. (2020c). *Overweight and Obesity.* Centers for Disease Control and Prevention. https://www.cdc.gov/obesity/index.html

National Center for Chronic Disease Prevention and Health Promotion. (2020d). *State and Local Programs.* https://www.cdc.gov/nccdphp/dnpao/state-local-programs/index.html

Parham-Payne, W. (2013). Weight perceptions and desired body size in a national sample of African American men and women with diabetes. *Journal of African American Studies, 17*(4), 433–443.

Piercy, K. L., & Troiano, R. P. (2018). Physical activity guidelines for Americans from the U.S. Department of Health and Human Services: Cardiovascular benefits and recommendations. *Circulation: Cardiovascular Quality and Outcomes, 11*(11).

Scarinci, I. C., Moore, A., Wynn-Wallace, T., Cherrington, A., Fouad, M., & Li, Y. (2014). A community-based, culturally relevant intervention to promote healthy eating and physical activity among middle-aged African American women in rural Alabama: Findings from a group randomized controlled trial. *Preventive Medicine, 69*, 13–20.

Sinclair, S., Hammond, D., & Goodman, S. (2013). Sociodemographic differences in the comprehension of nutritional labels on food products. *Journal of Nutrition Education and Behavior, 45*(6), 767–772.

Smith, S. L., Quandt, S. A., Arcury, T. A., Wetmore, L. K., Bell, R. A., & Vitolins, M. Z. (2006). Aging and eating in the rural, southern United States: Beliefs about salt and its effect on health. *Social Science & Medicine, 62*(1), 189–198.

Tumwebaze, J., Carter, V. L., & Dawkins, N. L. (2021). Enhancing the awareness of modifiable risk factors for cancer prevention in the Alabama black belt region: A follow-up study. *Journal of Health Care for the Poor and Underserved, 32*(4), 1995–2011.

University of Wisconsin Population Health Institute. (2014). *Alabama.* County Health Rankings and Roadmaps. http://www.countyhealthrankings.org/app/alabama/2014/measure/outcomes/1/map

University of Wisconsin Population Health Institute. (2021). *Alabama.* County Health Rankings and Roadmaps. https://www.countyhealthrankings.org/app/alabama/2021/rankings/bullock/county/outcomes/overall/snapshot

U.S. Department of Health and Human Services. (2018). *Current Guidelines.* Health.gov. https://health.gov/our-work/physical-activity/current-guidelines

Willett, W. C., & Stampfer, M. J. (2013). Current evidence on healthy eating. *Annual Review of Public Health, 34*, 77–95.

Woll, A., Jekauc, D., Niermann, C., & Reiner, M. (2013). Long-term health benefits of physical activity: A systematic review of longitudinal studies. *BMC Public Health, 13*(813). https://doi.org/10.1186/1471-2458-13-813

World Health Organization. (2021). *Obesity and overweight: Key facts.* World Health Organization. https://www.who.int/news-room/fact-sheets/detail/obesity-and-overweight.

JOURNAL OF

Extension

Research in Brief

Volume 60, Issue 2, 2022

Toward a More Effective Leader:
Planning for the Next Extension Administrator

KENNETH R. JONES[1]

AUTHOR1: [1]University of Kentucky.

Abstract. This study was conducted to assess the level at which state Cooperative Extension systems have strategies in place for administrative leadership changes. The data revealed that institutions have succession plans ranging from those that are very robust to very limited in nature. However, only 50% reported having individuals in key positions necessary to support continuity. In addition, 75% noted that it would take a year or more to replace the current Extension director/administrator if the person left immediately. This article provides insight on the successes and challenges associated with retaining top talent and mentoring potential leaders for advancement.

INTRODUCTION

The ever-changing administrative transitions among today's universities are viewed as a leadership crisis (Appadurai, 2009). Many are impacted by societal challenges, budget restraints, and other issues that perpetuate declining enrollments (Jaquette & Curs, 2015; Lamm & Israel, 2013). The Cooperative Extension Service has not gone unaffected. Hence, there is a need for strong leadership of state Cooperative Extension systems. There are fewer and fewer campus Extension administrators who matriculate from county and regional roles to specialists and associate deans/directors of Extension. Transition time for a predecessor to work closely with an incoming administrator to ensure continuity is limited. Moreover, additional present-day realities of limited resources, ongoing fiscal deficits, and changing political climates impact stakeholder interests (Monk et al., 2019; Page & Kern, 2018). As a result, Extension needs leaders with the capability to accept the challenges at hand (Berven et al., 2020; Godwin et al., 2011).

Literature on succession planning focuses primarily on business or corporate models, which limits concepts applicable to Extension systems (Lindner, 2001). Lindner (2001) argued that perhaps the most significant divergence between Extension and business is that while Extension's managers have been primarily internal hires or promotions, businesses rely on recruiting both internal and external professionals. Succession planning requires being aware of the major positions within an organization and setting forth action plans to prepare individuals to fill designated roles (Lindner, 2001; Luna, 2012; Day, 2007). This process broadens the talent pool, providing access to highly skilled employees who are prepared to assume positions as they become available. Succession planning has varied definitions across business and non-profits. There's even more ambiguity among units within the academy (Gonzalez, 2010). Croteau and Wolk (2010) emphasized a need for a paradigm shift: a push for leaders to think critically and execute ways to develop talent within the rank-and-file members of an organization. The benefit to creating a pipeline of future leaders from within is that employees are encouraged and empowered to develop as leaders, which fosters a sense of value and sparks a higher level of commitment.

Succession planning requires being proactive in preparing future leadership. It is not a threat to current administration, but rather it is an opportunity to perpetuate internal strengths. It is essential to prepare for the inevitable resignations or retirements (Bradley et al., 2012; Benge et al., 2015). Despite the intentional succession planning steps taken by corporations (and at times, government agencies), universities, colleges and particularly state Cooperative Extension systems seldom adopt similar practices (Lindner, 2001; Luna 2012; Wallin, 2007). This study was designed to analyze how leaders of state Extension systems internalize succession planning and to assess what plans are in place to nurture future leadership.

The purpose of the study was to assess the succession planning strategies among state Cooperative Extension Systems. The specific objectives were to:

1. Determine whether state Cooperative Extension Systems are preparing for administrative leadership changes.

2. Identify the top personal/core and technical competencies needed by Extension leaders.

3. Describe the institutional knowledge necessary to be successful as an Extension administrator.

4. Determine challenges and best practices associated with effective succession planning.

METHODS

The study included a convenience sample (Patton, 1990) of Extension leaders who voluntarily provided feedback in response to an online survey. The survey was designed to gather feedback from Extension administrators to assess whether state Cooperative Extension systems are preparing for administrative leadership changes. Extension leadership was defined as the individual directing Extension/land-grant operations. The titles varied among institutions (e.g., Extension Director, 1890 Administrator, Assistant/Associate Dean, Dean). Administrators were identified via purposive sampling approaches by reviewing a list of Extension administrators from all 50 states, compiled to identify those from each 1862 and 1890 land-grant university. Qualitative and quantitative data were collected. Narrative transcriptions from participant feedback (open-ended questions), multiple choice items, and responses to Likert-scale items from the online survey were analyzed using initial coding to identify primary themes. Participant responses were examined using a constant comparative approach (Creswell, 2003; Glasner & Strauss, 1967). In October of 2019, a total of 66 Administrators from 1862 and 1890 institutions in all 50 states were emailed a link to access the survey. A follow-up email was sent to non-responders in December and again in January 2020 in order to improve the response rate.

RESULTS

A total of 31 Extension administrators (47%) responded to the survey. Two of the respondents did not identify their state or institution. Responses from 25 states were included, representing the Extension regions (some states had both 1862 & 1890 land grants responding). The number of institutions by region were: Northeast (3); North Central (4); Southern (17; this includes 5 1890s); Western (5), and; unknown (2). All who responded were Extension administrators who were in charge of or had the influence to lead succession planning

efforts. The participants were asked a series of questions about their Extension system's administrative structure and succession planning efforts. When asked "Does your state Extension system have a plan to replace talented, highly valued administrators?", 50% responded "Yes", while 39% indicated "No" and 11% were unsure. Twenty-one percent reported that their succession plans do not begin until after the current administrator has announced plans to leave or retire.

When asked, "Who has the primary responsibility for succession planning in your Extension system?", most responses referenced those in upper-level administration, including Deans, Provosts/Vice Presidents and Chancellors/Presidents. However, when asked if they feel as though these individuals have the proficiency (e.g., resources, political acumen, etc.) to successfully select the right person or next leader for future leadership positions, 21% indicated feeling less assurance that this would occur.

Participants were also asked how long it would take to place a permanent hire (non-interim) in the position if the current Extension administrator left tomorrow. Seven individuals (23%) indicated that their Extension system is poised to replace the person within six months, while the remainder reported longer time frames. Approximately 60% noted that it would take about a year and 14% expressed that it would take longer than a year.

A primary objective of the study was to determine the status of land-grant institutions in regard to succession planning preparation. When asked if Extension systems were currently preparing a specific person to become the next administrative leader, only five reported using this strategy. Figure 1 shows that of the 28 responding to the survey item, a majority of institutions were not preparing their next Extension leader or were unsure whether this practice was occurring.

A total of 71% reported that their Extension system has a written document (e. g., job description, position expectations, etc.) that outlines the skills, competencies, and experiences expected of the next administrative leader. The remaining respondents (29%) indicated that there was no such document or that they were uncertain if one existed. For those (50%) reporting that plans were intact to replace administrators who leave, the level of depth varied. The quotes regarding plans included:

- "We make sure the necessary functions are covered from existing staff either by naming an interim or dispersing the duties. We then evaluate the position relative to current and future needs and make adjustments if necessary while it is open and then proceed. If it is necessary, we name a long term (plan on 1-2 years) interim while we conduct a search. The interim is often someone we have groomed for the position and is typically an internal candidate."

Toward a More Effective Leader: Planning for the Next Extension Administrator

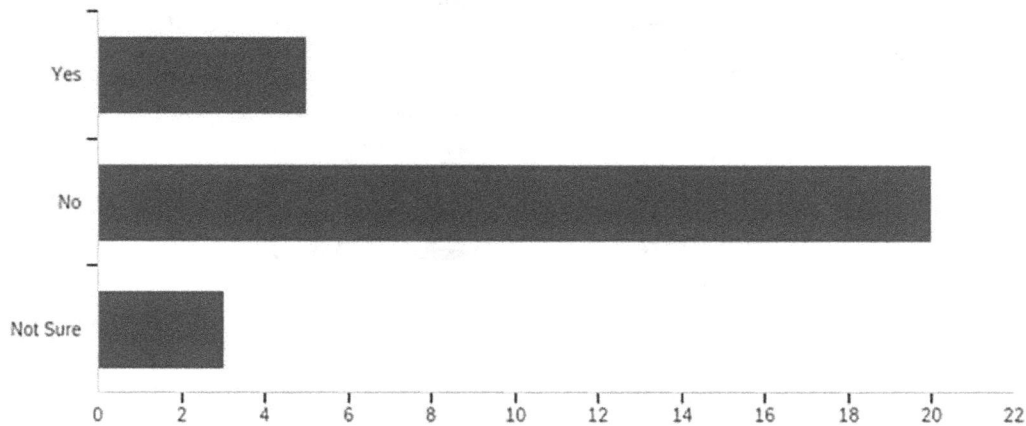

Figure 1. Institutions noting whether someone was in preparation of becoming the next administrative leader.

- "Certain individuals have been recruited and supported by administration in attending/participating in a variety of leadership opportunities and programs. Their leadership is valued and recognized by peers in state and nationally."

- "We advertise the position."

- "We typically pull from within our mid-management ranks (district, regional and unit leaders). Many of these individuals have been identified and have completed some advanced leadership and management training."

- "A group of talented professionals in programmatic leadership roles are currently in positions that groom them to assume administrative leadership roles as they become vacant."

Follow up questions asked participants to provide their perception of the competencies necessary to ensure leaders are successful. More specifically, the question was, "What top three personal/core competencies do you feel are important for an Extension Administrative Leader?" A list of competencies were provided and an additional option for "other" was included in order to give the chance to "write in" other competencies. A total of three competencies could be selected. The list of competencies provided within the survey was based on related work of several Extension scholars (Atiles, 2018; Berven et al., 2021; Scheer et al., 2011). Figure 2 reveals several competencies identified by those responding to the survey item.

Participants were also asked "What top three technical competencies do you feel are most relevant for an Extension Leader?" Up to three competencies could be selected, including an option to include "other" suggestions. Figure 3 shows the top technical competencies that were viewed as critical to Extension leaders. Note that in addition to the choices included in the survey, participants also provided additional examples (see "other comments") deemed as critical competencies.

Administrators were asked to respond to Likert-scale items to further assess their perceptions of succession planning within their Extension system. Table 1 summarizes those responses.

THE ROLE OF INSTITUTIONAL KNOWLEDGE IN 'HITTING THE GROUND RUNNING'

It should come as no surprise that most Extension administrators either want leaders who can maintain the status quo if the current conditions are favorable (at least at the beginning) or one who has the fortitude to make transformative changes that everyone can embrace. This is the expectation: for the new leader to 'hit the ground running'. In fact, this was evident in the findings of this study. Participants were asked to describe the institutional/organizational knowledge necessary to be successful as an Extension Leader. Responses included being knowledgeable of Extension; having the ability to lead/manage effectively, connect with staff, engage in lifelong experiential learning, and be a visionary; and having the wherewithal to step into/take on the role immediately. Given the common phrase that suggests Extension leaders must be able to hit the ground running, this study included a question to gather insight on this notion. More specifically, participants responded to a particular item: "It is often expressed that there is a need for a new administrator to *be ready now* or to *hit the ground running*. Describe what that means to you." The themes generated from responses related to institutional knowledge in relation to the need to hit the ground running are included in Table 2.

The administrators participating in the study perceived an array of challenges associated with succession planning (Table 3). Several noted that there was a lack of interest by

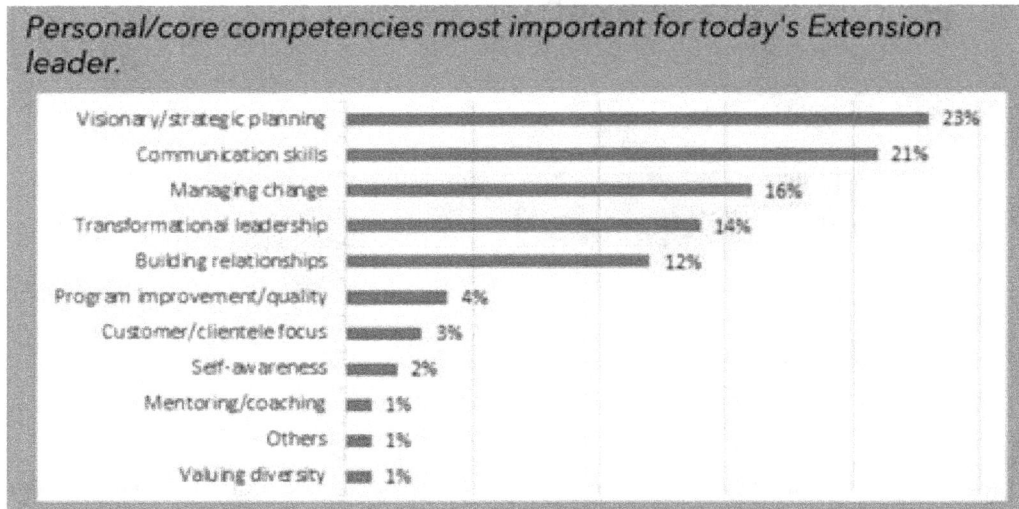

Figure 2. Core competencies as perceived by Extension administrative leaders.

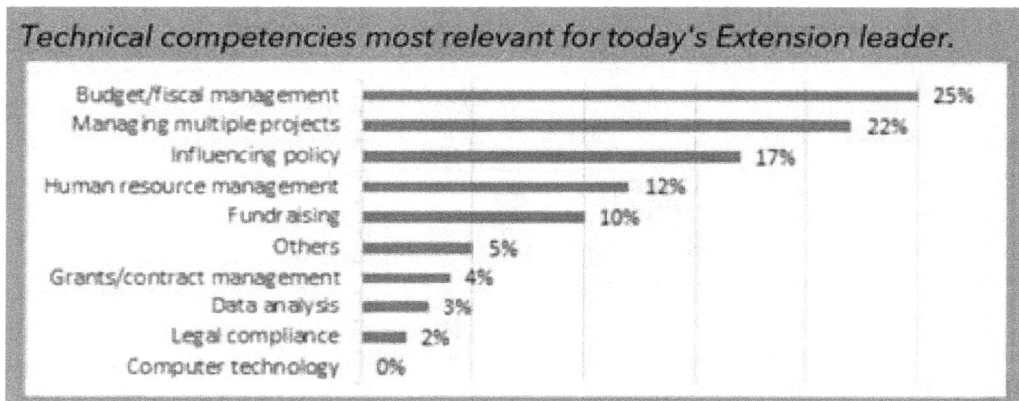

Figure 3. Technical competencies as perceived by Extension administrative leaders. "Other" comments from respondents included technical competencies such as clear communication of organizational vision, a historical perspective of Extension and program development, and an understanding of clientele needs.

Table 1. Extension Administrators' Perceptions of Succession Planning (Within Their Extension System)

#	Items	Mean	S.D.
1	I believe a new administrator must be able to hit the ground running in order to move Extension forward.	3.93	0.77
2	I believe I have a role to play in Extension's succession planning.	4.40	1.08
3	I am sought out for my opinion on strategies for succession planning.	4.13	0.72
4	I believe other administrators value my opinion on succession planning.	4.20	0.65
5	There is a high level of job satisfaction among those in position to move into administration.	3.93	0.68
6	There is a high level of engagement (serving on committees, volunteering to lead initiatives, etc.) among those in position to move into administration.	4.13	0.50
7	My Extension system is very effective in mentoring talented individuals for future administrative positions.	3.20	0.75

Note. Scale ranged from 1 (Strongly Disagree) to 5 (Strongly Agree).

Table 2. Extension Administrator Perceptions of the Notion to "Hit the Ground Running"

Theme	Example Quote
Connect with staff	A new administrator needs to be "ready" to listen, learn, and engage with staff, assess strengths, weaknesses and opportunities and then begin developing a plan to move forward.
Proper Management Skills	In order to "hit the ground running" a new administrator must already have the historical knowledge, self-awareness, emotional intelligence, strategic vision, planning and management skills to do the job immediately at the next level. Many of the skills at the "next level" are typically acquired through experience at the next level. In my opinion this is why succession planning is so difficult.
	An individual needs to be prepared to take on the day to day issues with little flexibility to learn as they go. Can they pick up the baton and carry on without missing too many beats?
Visionary	Come with the passion and a vision to enhance areas that are weak and determine what should not be done again.
Flexibility for Experiential Learning	It means they're willing to jump in, roll up their sleeves and get to work. They may not know everything, but they're willing to make decisions, make mistakes and try.
	I think it's an unrealistic comment. I agree that the individual needs to be ready to lead; however, you cannot fully embrace the job, until you have the job and that takes on-the-job training and support from others in the administrative team.
	I'm hesitant to subscribe to that philosophy. I think we are best served by an individual who feels challenged enough to spend time getting to know the organization from the vantage of their new position, whether they are an internal candidate or external. "Ready now" is fraught with potentially faulty assumptions and biases.

those currently in the system. Others reported that succession planning propels the belief that certain individuals are the 'favored ones' guaranteed to get all of the opportunities. Other reasons shared by the respondents included not having qualified individuals, human resource rules hindering the process, budget cuts, no priority on succession planning among decision makers, and a lack of opportunities/not enough positions available.

LIMITATIONS TO THE STUDY

This study was limited to only those who responded to the survey. Only 1862 and 1890 institutions were included in the study; therefore, 1994 land-grant institutions were not represented among the findings. There was also no comparison among regions or 1862/1890 universities, given the large number of participants among the Southern regions (which also included 1890 institutions in this study).

DISCUSSION

It is now more important than ever to institute plans for retaining valuable employees, including administrators. In some instances, we take for granted the talent that exists within our organizations or simply miss opportunities to prepare them for specific roles (Lindner, 2001). Other scenarios among Extension systems pertain to economics (Leuci,

Table 3. Reasons for Succession Planning Challenges

Question	% of Participants Responding	n
Lack of interest by current employee(s)	16%	5
Not having a qualified internal candidate	13%	4
Perceived favoritism of certain individuals within the system	16%	5
Other	45%	14

2012), thus requiring the time and effort to conduct searches for positions that could be filled by those currently among the ranks. When considering searching for talent from within, there may be the assumption that someone from the outside would be more qualified. It is often rare to find an external hire who has both the skills and institutional knowledge possessed by current employees. Another argument is that those who are already within the system are not fully prepared. If true, that is clearly a reason for succession planning. If a person has been productive in a middle management role, they should not be blamed for being unprepared to move to the next level. Perhaps they have not been given the opportunity for additional professional growth. Louder (2020) reported that individuals may lack the capacity for leadership simply because they have not had the chance to serve in roles that help further develop their skills.

Based on the findings presented, there is a broad range in which state Extension systems are engaged in succession planning. With only 50% responding that they have what they perceive to be a plan in place, it is unclear why others are not implementing a plan. Another question asked related to the number of times Extension administrators had been replaced within the last two years. Positions were filled with external hires/candidates from 0 to 7 times; positions were filled with internal hires/candidates from 0 to 12 times. Since internal hires are made more frequently, it is important to ensure that these individuals are prepared with the skills necessary for upward mobility.

Institutional knowledge can often be drastically undervalued within academia (Duderstadt, 2007). The majority of administrators participating in this study emphasized the importance of having knowledge and expertise with land-grant systems/Extension, as well as an understanding of how to build partnerships and influence policy within higher education. It is important to note that some of the concepts described cannot be formally taught, only gained through years of experience (Seger & Hill, 2016). Arguably, succession planning can enable Extension systems to prepare individuals for leadership opportunities. It is unrealistic to expect one with few opportunities for professional growth to take on major leadership responsibilities with no challenges (Griffeth et al., 2018; Louder, 2020). Sharing knowledge through informal mentoring or documented efforts is helpful not only to new leaders, but for the organization as a whole.

When asked what administrative leaders perceived to be the most challenging issue in regard to succession planning within Extension, common themes centered on the lack of qualified individuals (see Table 1). This issue can be minimized by investing in employees to help them develop the wherewithal to lead. A few leadership programs and initiatives exist for Extension professionals, including LEAD 21 and the Food Systems Leadership Institute (FSLI). These provide intensive professional development specifically for the next generation of land-grant administrators. Another common expression that stifles considerations for succession planning is the perceived favoritism. When there is a perception among the organization that a lot of meaningful work assignments tend to go to a select few who are most visible and at times most vocal, a morale issue can emerge. Current leaders should make sure both equity and equality are considered and use discernment when identifying the most qualified and sharing options for employees to grow.

Within the Extension landscape, there are obviously prevailing threats as previously discussed, but also new opportunities. Institutions must take a serious look at succession planning as an essential element for cultivating leadership.

Current Extension leaders could build human capital from within while enhancing the organization at multiple levels by targeting aspiring administrators, such as middle managers and unit directors. This, in turn, helps to develop a group of what Wallin (2007) refers to as a core of well-informed, qualified, supportive people who understand the institution. This is indeed an effective way to foster better leadership and relationships from a systematic approach.

A few recommendations are provided below, based on the data examined for this study and from similar work that relates to succession planning. The aim is to help Extension administrators consider best practices to foster leadership development for those with the potential to be future Extension leaders (particularly those interested in campus level administration).

- Seek out a number of potential leaders and not just one or two individuals. This can minimize the perception of favoritism; namely that multiple people may be presumed worthy for advancement.

- Promote self-awareness among those who may be interested in becoming Extension administrators by exposing them to projects that require the use of their abilities. It may be best to allow them to identify their own weaknesses before others offer critiques. Be sure to use this as a coaching opportunity while taking note of their strengths. Although they may not currently be ready for certain leadership roles, continue to monitor progress in consideration for future opportunities.

- Job-shadowing can exist at all levels. Offer employees of promise a chance to join in at meetings where decisions are made. Note their comfort level and ability to offer feedback or suggestions.

- Resist the temptation to micro-manage. Effective administrators cannot sacrifice time on this (if you find yourself constantly doing so, this may be a clue that you need to move on to groom another/others). This is not to say that you should give up on those who don't immediately meet expectations. Make it your responsibility to identify talents of employees, then make assignments accordingly. Acknowledge that some people don't know their own strengths until others help identify what they bring to the table.

- Take heed to the specific core and technical competencies that have been identified by other Extension scholars as critical to today's Extension professional. Many of those same competencies are mentioned in this study. Assisting future Extension leaders in further developing their abilities in these areas could help expedite their route to an administrative position.

- Selecting, hiring, and promoting individuals are only parts of effective succession planning. Even the best hires require some mentoring. Table 1 (item #7) denotes that participants in this study had less than positive perceptions of existing mentoring practices. Reviewing recent literature for practical strategies on being an effective mentor may prove beneficial.

- Look beyond typical inner circles. Future leaders come from all genders, races and cultures. From a Cooperative Extension lens, there is also a need to be inclusive by considering individuals from all program areas. Historically, the leaders of Extension systems have been predominantly from the traditional areas of the agricultural sciences (e. g., animal, food, plant/soil, natural resources, etc.). Selecting primarily from one group of applicants is an inequitable approach. Furthermore, it ignores the fact that many of today's issues in need of Extension's attention deal abundantly with the societal ills and dysfunction among families and communities. In essence, we need leaders with expertise in the social science side of agriculture as much as we need those who understand the dynamics impacting production agriculture within our states and country. When diversity is fully embraced, not only does it help to nurture a better appreciation for those we are leading, but it helps leaders become more perceptive in addressing personal limitations (Griffeth et al., 2018). Leaders set in their comfort zones can impede vision; moving beyond that proverbial barrier can serve as a personal growth machine.

SUMMARY

Cooperative Extension could benefit from strategies that integrate leadership development into its succession planning efforts. Succession planning is not simply replacing current leaders. In addition, solely relying on institutional wisdom without proper planning is problematic. State Extension systems should create environments where staff and faculty are encouraged to pursue leadership opportunities with confidence. Succession plans should be clearly communicated as an intentional modality that is embraced by upper-level administration in charge of promotional processes at the college level and beyond. Turnover in leadership is inevitable, from resignations to retirements. The key is to prepare for replacements in a strategic manner. Future studies are needed to more closely examine how succession planning fits effectively within complex Extension systems. However, being proactive versus reactive is the initial premise.

REFERENCES

Appadurai, A. (2009). Higher education's coming leadership crisis. *The Chronicle of Higher Education, 55*(31), A60. https://eric.ed.gov/?id=EJ838831

Atiles, J. (2018). Cooperative Extension competencies for the community engagement professional. *Journal of Higher Education Outreach and Engagement, 23*(1), p. 107–127. https://files.eric.ed.gov/fulltext/EJ1212491.pdf

Benge, M., Harder, A., & Goodwin, J. (2015). Solutions to burnout and retention as perceived by county Extension agents of the Colorado State University Extension System. *Journal of Human Sciences and Extension, 3*(1). https://www.jhseonline.com/article/view/606/531

Berven, B. C., Franck, K. L., & Hastings, S. W. (2021). Investing in Extension's workforce: Assessing and developing critical competencies of new agents. *Journal of Extension, 58*(2). Available at: https://tigerprints.clemson.edu/joe/vol58/iss2/28

Bradley, L., Driscoll, E., & Barden, R. (2012). Removing the tension from Extension. *Journal of Extension, 50*(2). Available at: https://archives.joe.org/joe/2012april/tt1.php

Creswell, J.W. (2003). *Research design: Qualitative, quantitative and mixed method approaches.* (2nd ed.). Sage.

Croteau, J. D. & Wolk, H. G. (2010). Defining advancement career paths and succession plans: Critical human capital retention strategies for high performing advancement divisions. *International Journal of Educational Advancement, 10*(2), 59–70. https://eric.ed.gov/?id=EJ907750

Day, D. V. (2007). *Developing leadership talent: A guide to succession planning and leadership development.* Society for Human Resource Management. https://www.shrm.org/hr-today/trends-and-forecasting/special-reports-and-expert-views/Documents/Developing-Leadership-Talent.pdf

Duderstadt, J. (2007). *The view from the helm: Leading the American University during an era of change.* University of Michigan Press.

Glasner, B. G. & Strauss, A. L. (1967). *The discovery of grounded theory: Strategies for qualitative research.* Aldine de Gruyter.

Godwin, D., Diem, K. & Maddy, D. (2011). Best management practices for a successful transition into an administrative role. *Journal of Extension, 49*(4). https://archives.joe.org/joe/2011august/a1.php

Gonzalez, C. (2010). *Leadership, diversity and succession planning in academia.* Research and Occasional Paper Series, Center for Studies in Higher Education, UC Berkeley. https://files.eric.ed.gov/fulltext/ED512031.pdf

Griffeth, L. L., Tiller, L., Jordan, J., Sapp, R., Randall, N. (2018). Women leaders in agriculture: Data-driven recommendations for action and perspectives on furthering the conservation. *Journal of Extension, 56*(7). https://archives.joe.org/joe/2018december/a2.php

Jaquette, O. & Curs, B. R. (2015). Creating the out-of-state university: Do public universities increase nonresident freshman enrollment in response to declining state appropriations? *Research in Higher Education, 56*, 535–565. www.doi.org/10.1007/s11162-015-9362-2

Lamm, A. & Israel, G. (2013). A national examination of Extension professionals' use of evaluation: Does intended use improve effort? *Journal of Human Sciences and Extension, 3*(1). https://www.jhseonline.com/article/view/657/568

Leuci, M. S. (2012). The nature of organizational learning in a state Extension organization. *Journal of Extension, 50*(3). https://tigerprints.clemson.edu/joe/vol50/iss3/35

Lindner, J. R. (2001). Competency assessment and human resource management performance of county Extension chairs. *Journal of Southern Agricultural Education Research, 51*(1). http://jsaer.org/pdf/Vol51/51-00-333.pdf

Louder, E. R. (2020). *Leadership styles and barriers to leadership for women in agriculture: A mixed methods study.* (Publication No. 7996) [Graduate Thesis, Utah State University] All Graduate Theses and Dissertations. https://digitalcommons.usu.edu/etd/7996

Luna, G. (2012). Planning for an American higher education leadership crisis: the Succession issues for administration. *International Leadership Journal, 4*(1). https://citeseerx.ist.psu.edu/viewdoc/download?doi=10.1.1.464.5025&rep=rep1&type=pdf

Monk, J. K., Vennum, A. V., & Kanter, J. B. (2019). How to use crowdfunding in Extension: A relationship education example. *Journal of Extension, 57*(4). https://archives.joe.org/joe/2019august/tt3.php

Page, C. S., & Kern, M. A. (2018). Creating and implementing diverse development strategies to support Extension centers and programs. *Journal of Extension, 56*(1). https://archives.joe.org/joe/2018february/a4.php

Patton, M.Q. (1990). *Qualitative evaluation and research methods* (2nd ed.). Sage.

Seger, J., & Hill, P. (2016). The future of Extension leadership is soft leadership. *Journal of Extension, 54*(5). https://tigerprints.clemson.edu/joe/vol54/iss5/21

Scheer, S. D., Cochran, G. R., Harder, A., & Place, N. T. (2011). Competency modeling in Extension education: Integrating an academic Extension education model with an Extension human resource management model. *Journal of Agricultural Education, 52*(3), 64-74. https://doi.org/10.5032/jae.2011.03064

Wallin, D. (2007). Succession planning: Developing future leaders from within. *Academic Leader, 23*(1), 2–8.

JOURNAL OF
Extension

Feature Article

Volume 60, Issue 2, 2022

The Adoption of Food Safety Practices and the Implications of Regulation for Small-Scale Farms

ELIZABETH CANALES[1], JUAN SILVA[1], AND JOY ANDERSON[1]

AUTHORS: [1]Mississippi State University.

Abstract. In this article we examine the adoption of food safety practices among produce growers in the south and discuss implications of food safety regulations in the U.S. Produce growers have adopted standard food safety practices to varying degrees, but there is still an adoption gap, particularly among small scale operations. Market-driven and regulatory food safety enforcement continues to tighten, and this can further hinder market access for small scale producers.

INTRODUCTION

Fresh produce has often been linked to cases of foodborne illness, making food safety a top priority for the produce industry due to the economic and public health impacts (IFSAC, 2018). The food industry has adopted various private food safety standards to manage risks in the supply chain for fresh produce. For producers to gain access to larger markets such as wholesale, foodservice, and retail, they generally need food safety certifications, third-party audits, or food safety trainings (e.g., Produce Safety Alliance food safety training). Buyer enforcement of food safety standards varies, with some markets imposing stricter standards than others. Some examples of third-party audits or certifications are the Good Agricultural Practices (GAP) certification (e.g., USDA GAP, Global G.A.P.) and the California Leafy Green Marketing Agreement (LGMA).

In addition to market-driven initiatives, the Produce Safety Rule (PSR) of the Food Safety Modernization Act (FSMA) gives the FDA authority to regulate fresh produce. Under this rule, the FDA has issued food safety standards for the growing, harvesting, packing, and holding of fresh produce with compliance dates beginning in 2018 for large farms (FDA, 2020a). FSMA shifts the focus from simply responding to food safety issues to preventing them starting at the farm level. Most small and medium scale farmers selling directly to consumers are exempt from the PSR. However, that does not preclude buyers from requesting adherence to the practices included in this rule. Anecdotal evidence suggests that some buyers have started requesting growers to adhere to the minimum food safety standards included in the PSR. The lack of adherence to food safety programs can limit grower access to markets, which could primarily impact small-scale producers who lack the capital and infrastructure to adopt practices at the level demanded by some markets.

Some direct-to-consumer outlets do not regulate or require any third-party food safety audits for producers. A study by Harrison et al. (2013) found that few farmers markets request food safety information from vendors. For many, the fact that the PSR exempts most small- and medium-scale farms means these farmers are left with no food safety inspection. The adoption of food safety practices can be costly and given that no audit or certification is required by some outlets, there are fewer economic incentives for small-scale producers to pursue certifications and invest in third-party audits. Although most food safety incidents are associated with large-scale operations, industry leaders representing large-scale interests are concerned about the safety of food produced by small-scale producers that sell through direct-to-consumer market channels (Parker et al., 2016). As direct-to-consumer channels continue to grow in importance, there will be pressure for these channels to provide improved food safety assurances.

In this article, we review the adoption of on-farm food safety practices with a focus on small-scale producers and operations with direct-to-consumer sales. We discuss some insights about producers' views and perceived barriers as well as implications of tighter food safety regulations for small produce operations.

METHODS AND RESULTS

We report the adoption of food safety practices using data from a fruit and vegetable growers survey. The survey was administered during in-person Extension food safety workshops in Mississippi in 2018 and 2019. Growers were asked to complete the survey at the beginning of the workshop. The link to an online version of the survey in Qualtrics was also shared with other growers following a snowball sampling approach. The practices examined are based on standard GAP that are part of the requirements in the PSR. The sample (n=79) consists of growers from Mississippi, Alabama, and Arkansas, making the sample more representative of the deep South. Fifty-six percent of the growers had average annual sales less than $25,000 and 21% had sales between $25,000–$100,000. Overall, 77% of the operations in the survey are considered low-sales small farms (i.e., farm with sales of less than $100,000 (USDA, 2019)). On average, the share of direct-to-consumer sales was 62%, with more than half of the operations having a share of direct-to-consumer sales above 80%. Thus, this sample is weighted toward small-scale farms selling through direct-to-consumer market channels.

Figure 1 depicts the level of adoption of food safety practices for small scale operations (<$100,000 revenue) and for operations with a large share (50% or more) of direct-to-consumer sales. We then examine the association between farm characteristics and the adoption of on-farm food safety practices using logit regression, an approach commonly used to model binary dependent variables. The factors we examine are farm size (log of fruit and vegetable acreage), the share of direct-to-consumer sales, whether the operation grows leafy greens, and if the operation is organic/sustainable (this category includes designations of certified organic, in transition to certified organic, and sustainably or naturally grown certified operations). We include these variables because there are concerns among stakeholder groups that adoption of food safety practices in the PSR and other food safety programs could be onerous for small-scale and organic or sustainable operations (Adalja and Lichtenberg, 2018a). Because the parameters from a logit regression are not easily interpreted, we report the marginal effects on the probability of adoption of these practices in Table 1. These marginal effects represent the change in the probability that a farmer would use or adopt a particular practice given a one unit change in the independent variables.

As observed in Figure 1, basic food safety practices have been adopted to various degrees by produce growers, yet there is a still a significant gap in the adoption of standard practices. Similar adoption gaps are reported by Adalja and Lichtenberg (2018a). The adoption of these practices is generally lower among small farms and operations with a higher share of direct-to-consumer sales. Only 56% of small operations provide food safety training to their employees, and 61% provide equipped toilets and hand washing stations. Like the adoption of other agricultural practices, size has a positive correlation with practice use (Table 1). A 1% increase

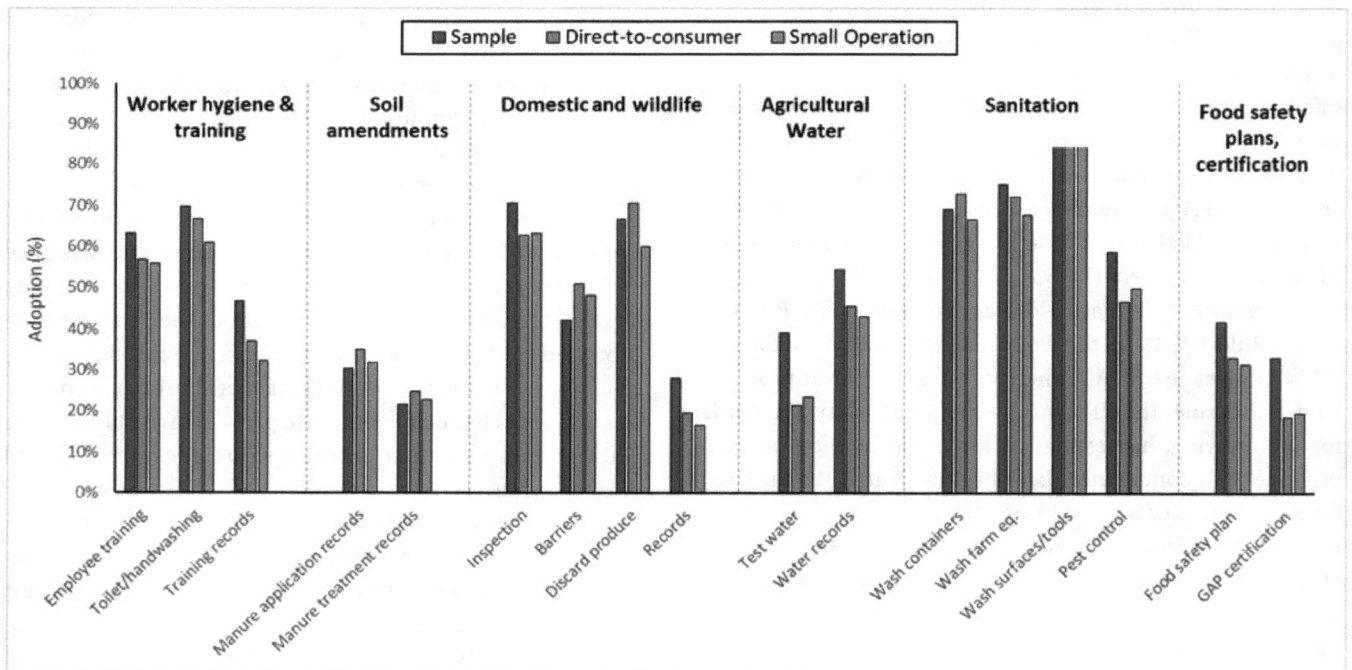

Figure 1. On-farm use of conservation practices.. Small operations are low sales farms with <$100,000 in revenue, and direct-to-consumer are operations with a share of direct-to-consumer sales of 50% or more.

Table 1. Marginal Effects on Probability of Practice Use

	Worker hygiene & training			Domestic and wildlife				Ag. water		Sanitation			Plans and cert.		
	Employee training	Toilet/ handwashing	Training records	Inspection	Barriers	Discard produce	Records	Water testing	Water records	Wash containers	Wash farm eq.	Wash surfaces, tools	Pest control	Food safety plan	Certified/ third-party audit
Log Acreage[1]	0.064* (0.027)	0.065* (0.028)	0.048* (0.027)	0.028 (0.029)	-0.005 (0.031)	0.036 (0.027)	0.037 (0.024)	0.088* (0.026)	0.049* (0.026)	0.018 (0.024)	0.020 (0.027)	0.047* (0.024)	0.082* (0.027)	0.043 (0.028)	0.043* (0.023)
Organic/ sustainable[2]	-0.158 (0.123)	0.038 (0.111)	0.016 (0.125)	0.001 (0.118)	-0.040 (0.136)	-0.282* (0.125)	0.031 (0.130)	-0.305* (0.125)	-0.283* (0.125)	-0.197 (0.142)	0.278** (0.117)	0.000* (0.000)	-0.258* (0.116)	-0.211 (0.131)	-0.013 (0.119)
Share of direct-to-consumer sales[1]	-0.001 (0.001)	-0.002 (0.001)	-0.003* (0.001)	-0.002 (0.001)	0.002 (0.001)	0.000 (0.001)	-0.002 (0.001)	-0.003* (0.001)	-0.005* (0.001)	0.001 (0.001)	-0.002 (0.001)	0.000 (0.001)	-0.003* (0.001)	-0.004* (0.001)	-0.004* (0.001)
Leafy Greens[2]	0.127 (0.119)	-0.012 (0.106)	0.118 (0.121)	-0.043 (0.113)	0.017 (0.130)	0.028 (0.121)	0.043 (0.120)	-0.122 (0.114)	0.258* (0.116)	0.365* (0.121)	-0.190* (0.109)	0.160 (0.122)	0.068 (0.119)	0.000 (0.129)	0.120 (0.114)
N	74	74	74	73	73	73	73	46	66	69	68	46	69	69	68

Note. The number of observations vary due to the number of missing values.

**, * indicates statistical significance at the 5% and 10% significance level, respectively. Standard errors are in parenthesis.

[1] Values can be interpreted as the percentage point change in the likelihood of adoption given a one percentage change in acreage or one percentage point increase in the share of direct-to-consumer sales.

[2] Values can be interpreted as the percentage point change in the likelihood of adoption for organic/sustainable farms compared to conventional farms, or for leafy green growers compared to growers who do not grow leafy greens, respectively.

in acreage is associated with a 6 percentage point increase in the likelihood that an operation provides hygiene and food safety training to its employees and a 5 percentage point increase in the probability of keeping training records (Table 1). While operations may provide some hygiene training, not all of them keep records of those trainings. We found that farms with a larger share of direct-to-consumer sales are less likely to keep training records. Wholesale and retail buyers generally require a third-party audit, on-site visits, or food safety records from the farm, and those requirements are not as prevalent in direct-to-consumer market channels (Harrison et al., 2013).

In our sample, only 23 respondents reported using animal-based soil amendments, and of those, only 30% and 22% keep records of manure treatment or application dates, respectively (Figure 1). Due of the low number of respondents using soil amendments, we did not include these practices in the logit regression analysis reported in Table 1. Inspection for domestic and wildlife intrusion was 63%, and the use of barriers to prevent contamination was 51%, for small operations. As shown in Figure 1, a higher percent reported discarding produce suspected of contamination due to animal contact (71%). We did not find statistical differences in the use of these practices across the different types of operations (Table 1).

Two of the practices with the lowest levels of adoption were related to water testing and recordkeeping, with 21% and 24% use respectively among small farms and operations with a large share of direct-to-consumer sales (Figure 1). Agricultural water is an important risk factor in the produce industry, as it has been identified as the source of contamination in multiple high-profile outbreaks (FDA, 2019; Cooley et al., 2007). We found that farm size is associated with an increase in the probability of water testing and keeping test records—a 1% increase in acreage is associated with an 8 percentage points increase in the likelihood of water testing and a 5 percentage point increase in the likelihood of recordkeeping (Table 1). We observe a negative association pattern between practice usage and direct-to-consumer sales for water testing and water recordkeeping. Some small-scale operations use municipal water or public water systems and may not need to test their agricultural water. However, while these growers receive or can request public water test reports, many do not maintain these records. As expected, due to compliance with some organic certification requirements, organic producers are more likely to test their agricultural water and maintain test records (Table 1). Operations that grow leafy greens are also more likely to keep water test records. Leafy greens have been associated with large, publicized foodborne outbreaks linked to agricultural water (Marshall et al., 2020) which may be why we see this result.

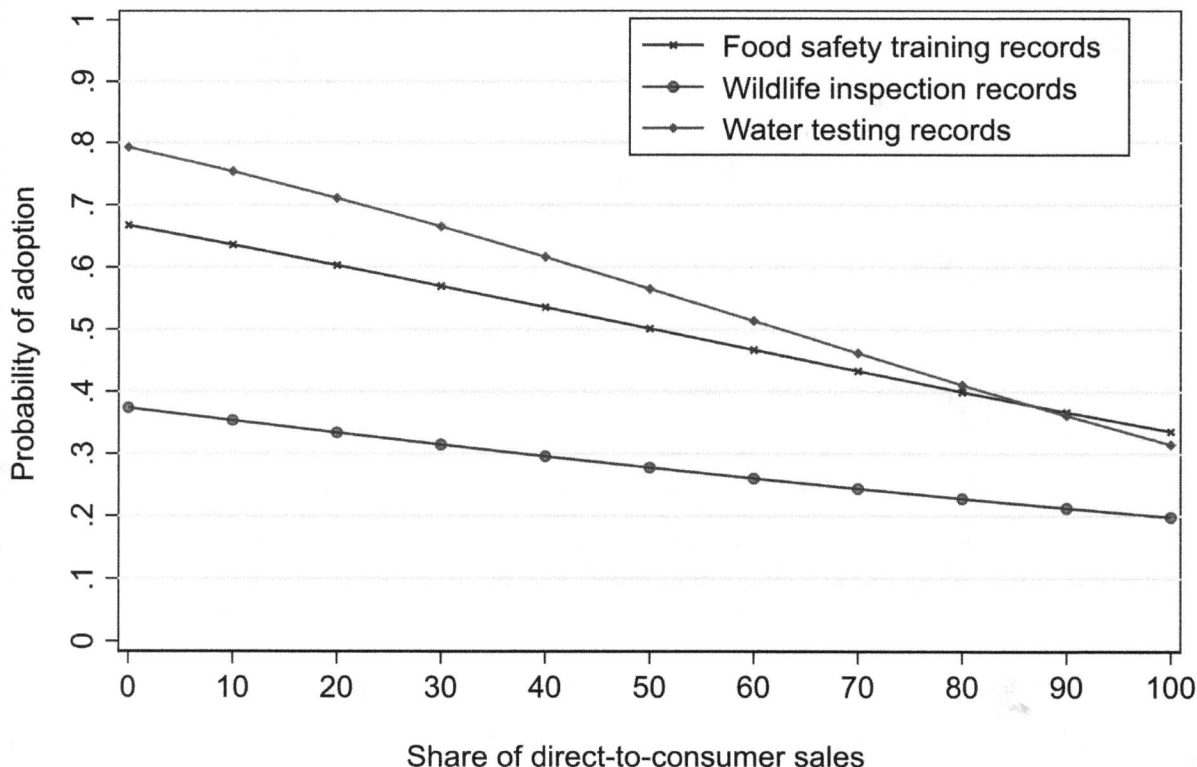

Figure 2. Probability of keeping food safety records for producers selling through direct-to-consumer outlets.

Larger operations are also more likely to wash surfaces and equipment that come in contact with produce. The likelihood of doing so increases by around 4 percentage points for each 1% increase in acreage (Table 1). Operations with leafy greens are also more likely to sanitize harvest containers, which is likely due to the risk associated with leafy greens. Pest control programs for packing or storage buildings are more likely to be adopted by larger and/or organic operations and less likely to be adopted by operations with a larger share of direct-to-consumer sales.

Implementing food safety practices can be costly, and farm size plays a significant role in the dynamics of adoption. Compliance with private or regulatory food safety standards could be burdensome, particularly for small-scale farms. Previous studies have found that per-acre food safety costs decrease as farm size increases, indicative of the presence of economies of scale (Adalja and Lichtenberg, 2018a; Hardesty and Kusunose, 2009).

FOOD SAFETY RECORDS

It is important to note that while producers may have adopted some food safety practices, the measures implemented may not be as robust as necessary to pass a farm audit or inspection. For example, food safety recordkeeping is lacking, as our results suggest. While producers use some practices, they are less likely to keep records to demonstrate compliance. As is seen in Figures 1 and 2, this issue is more prevalent among small-scale operations and producers selling through direct-to-consumer market channels.

Records are a key component of any regulatory and private food safety program. Records are key in demonstrating food safety efforts and compliance and in tracing outbreaks. The lack of records also makes traceback and root cause analysis difficult. From a risk management perspective, it is vital to communicate to producers the importance of maintaining good, sortable records. For many small growers selling directly to consumers, the lack of market enforcement provides no incentive to maintain records of food safety practices, yet documenting food safety measures can help protect producers from potential liability. As the industry increases efforts to improve traceability and the quality of records maintained along the supply chain, there will be a greater push for producers to improve their records and to move from a paper-based system to a sortable electronic-based system in order to improve the speed of tracing and identification of potential safety issues. For shorter supply chains (e.g., direct-to-consumer or retailers working within a state), tracing issues could be dealt with faster due to the simplicity of the supply chain. However, buyers and consumers could demand that small producers provide better food safety and traceability assurances similar to those of the broader food supply chain.

BARRIERS AND DRIVERS TO ADOPTION OF FOOD SAFETY PRACTICES

When we evaluate responses regarding farmers' motivations for not pursuing a GAP certification or third-party food safety audit, we find that the scale of production and the lack of market enforcement are the main forces at work. Many small-scale growers also perceive that there are low returns to pursuing a certification. For example, 45% of respondents in our survey disagree that having a food safety certification pays off and around 60% believe that food safety certification should not be required for small scale producers. Increasing growers' awareness of risks and the benefits of prevention is important from an economic point of view, as the cost of an outbreak far surpasses the cost of prevention (Ribera et al., 2012).

Farmers were asked to rank perceived barriers to the adoption of on-farm food safety practices. Figure 3 shows the percentage of respondents who ranked the listed barriers as the first or second main challenges to implementing food safety practices. Economic factors such as implementation cost and a lack of resources were reported as the main limiting factors. Studies have found that expenditures on food safety practices can be more burdensome for small producers due to economies of scale (Adalja and Lichtenberg, 2018a). Time constraints also play a role and may be an issue particularly for small- and medium-scale operations that do not have a dedicated professional food safety staff, as food safety tends to be juggled alongside other farming tasks. Limited knowledge was not often ranked as the main barrier to adoption but was ranked second by several producers. In general, the problem does not seem to be the lack of awareness of food safety concepts (Parker et al., 2016), but resource constraints and cost barriers that inhibit implementing food safety practices.

Beyond certifications, the adoption of individual food safety practices among non-certified operations is also largely driven by buyers' requirements (Figure 4). Although small scale producers may not need to be certified when selling through local or direct-to-consumer market channels, the lack of a certification or third-party audit significantly limits their access to new markets (Figure 4), and this may hinder their ability to grow their operation and expand sales.

DISCUSSION

THE FUTURE OF FOOD SAFETY ENFORCEMENT AND REGULATION

In 2020, the FDA published a footprint for a "New Era of Smarter Food Safety," which outlines the FDA's future approach to food safety. This new approach puts emphasis on the promotion of a food safety culture, better recordkeeping and traceability, and technology-enabled food safety systems (FDA, 2020d). The goal of this plan is to use modern approaches (e.g., technological tools, analytical techniques)

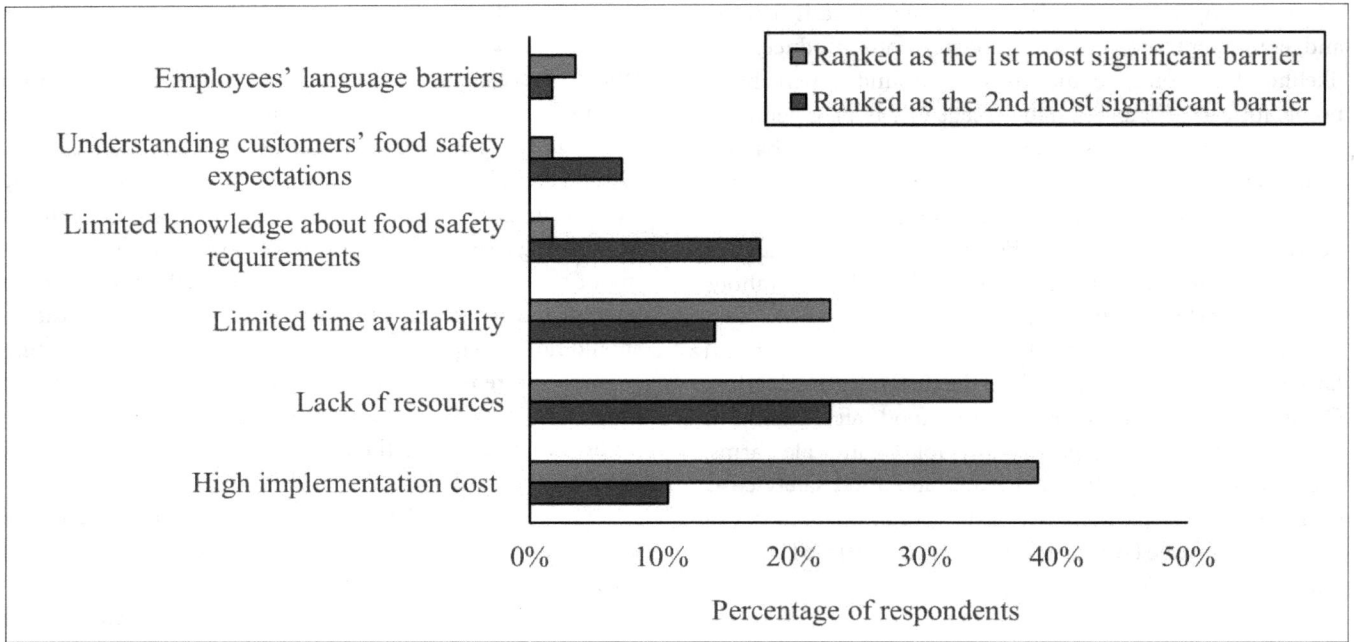

Figure 3. Ranking of perceived barriers to the adoption of on-farm food safety practices.

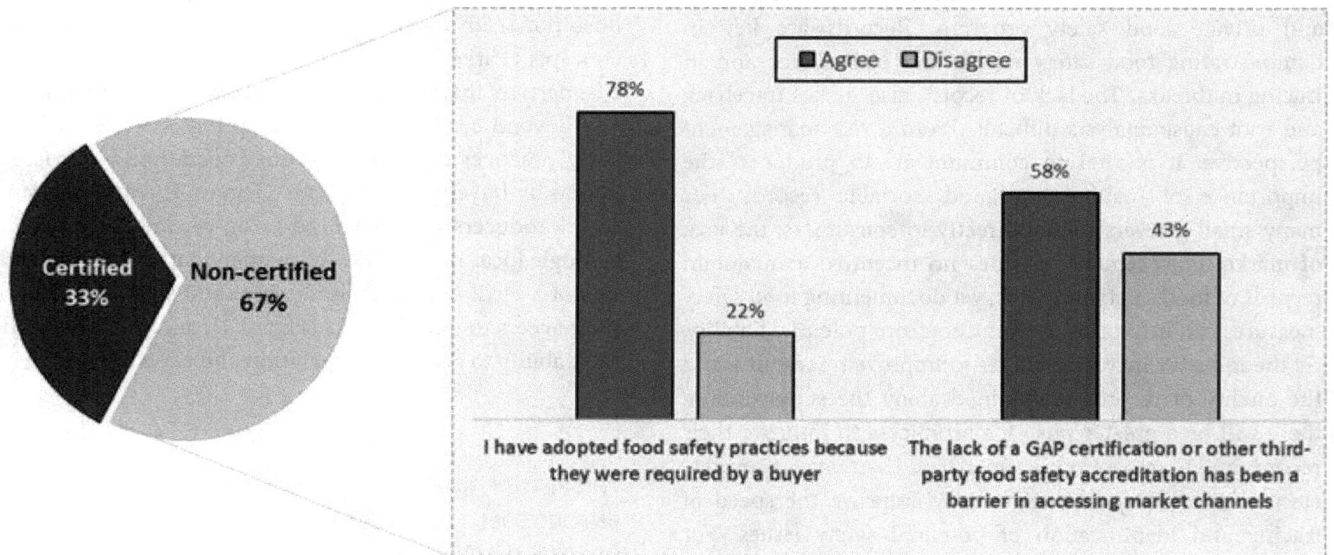

Figure 4. Food safety certification and market access.

to allow for real-time visibility of the food supply chain, focus on prevention, and improve the ability of the industry to identify and predict issues.

In 2020, the FDA also proposed the "Requirements for Additional Traceability Records for Certain Foods" rule as part of FSMA. The proposed rule was published in the Federal Register in September 2020 with the public comment period ending in February 2021. The compliance date is expected to be two years after the final regulation is effective (FDA, 2020c.). This rule heightens traceability recordkeeping requirements for operations that manufacture, process, pack, or hold high-risk products. Some examples of produce on the proposed rule's Food Traceability List include cucumbers, fresh herbs, melons, sprouts, tomatoes, leafy greens, tropical tree fruit, and fresh-cut fruits and vegetables (FDA, 2020b). While this proposed regulation does not rule out paper records, companies within the supply chain need to be able to provide electronic spreadsheets containing necessary traceability information (e.g., lot codes, receiving and shipping dates, and information regarding internal traceability programs) to allow for a rapid trace of products throughout the supply chain. Because of the lower volume produced by very small farms (i.e., farms with annual produce sales lower than $25,000) and farms selling directly to consumers, these operations will be exempt from the recordkeeping requirements applicable to food growers (FDA, 2020b). But if they are not exempt, subsequent entities buying from these small farms would still need to keep records of the foods produced by these operations. Thus, it is possible that small farms will face pressure to comply with stricter food safety and traceability recordkeeping requirements. There are also industry-led efforts to improve traceability within the produce supply chain. For example, the LGMA requires its members to have a traceability system in place and continues to tighten food safety requirements (Horsfall, 2020; Ward, 2020).

As the regulatory examples and the industry trends discussed above demonstrate, food safety regulations and private standards are not likely to ease but are likely to become more rigorous. The produce industry will continue to take steps to prevent foodborne outbreaks and ensure the safety of the supply chain. The adoption of elevated standards and emerging technological tools will likely impose structural barriers for small and medium scale producers for which the needed capital investment could be overly burdensome and unjustified given their scale of operation, the requirements, and the lesser complexity and risks of more localized markets. Yet as the industry moves in that direction, there is a need to develop scale-appropriate approaches that small-scale operations can adopt to make progress towards new industry demands.

IMPLICATIONS AND OUTREACH NEEDS

Increasing regulatory oversight is concerning to many produce stakeholders. A 2019 state of the vegetable industry survey reported that food safety rules are a main concern for 40% of produce stakeholders (Miller, 2019). The wider adoption of food safety private standards and the move towards more rigorous requirements and regulations can impact small and medium-scale farmers for whom it may be more difficult to access markets, as these practices and audits are more widespread and enforced. For example, a study simulating the effect of FSMA suggests that this program could result in market share losses for small growers that have to incur additional costs to comply with the rule (Bovay and Sumner, 2018). As regulatory agencies, buyers, and consumers demand higher levels of food safety compliance and transparency, some local market outlets (e.g., farmers markets) may also increase their food safety efforts by requiring food safety information and training from vendors. It is important to develop scale-appropriate approaches to food safety and education programs (Parker et al., 2016) to ensure the food produced and sold by local small-scale producers is safe in a way that allows these producers and markets to stay competitive.

Insights from our survey and other published articles (Rodrigues at al., 2020; Strohbehn et al., 2018) suggest that farmers need additional assistance in translating regulations into actionable items, navigating the food safety requirements of different buyers and markets, adopting practical tools, and receiving hands-on training to develop their food safety program. Recordkeeping was identified as an area where additional effort is needed. Education covering risk analysis and the value of maintaining good records beyond regulatory compliance is important. The adoption of scale-appropriate electronic recordkeeping (e.g., Excel-based templates) could be encouraged, and hands-on training could be provided to develop or adapt existing tools to the specific needs of each farm. Education involving peer-to-peer learning opportunities and on-farm site visits is needed to help producers learn scale-friendly cost-effective ways to improve food safety. Engaging local markets (e.g., managers of farmers markets) in training efforts and food safety initiatives could also be beneficial, as these outlets would also benefit from a reduction in food safety risks (Harrison et al., 2013).

REFERENCES

Adalja, A., & Lichtenberg, E. (2018a). Produce growers' cost of complying with the Food Safety Modernization Act. *Food Policy, 74*(C), 23-38. www.doi.org/10.1016/j.foodpol.2017.10.005

Adalja, A., & Lichtenberg, E. (2018b). Implementation challenges of the food safety modernization act: Evidence from a national survey of produce growers. *Food Con-*

trol, *89*, 62-71. https://doi.org/10.1016/j.foodcont.2018.01.024

Bovay, J., & Sumner, D. A. (2018). Economic Effects of the U.S. Food Safety Modernization Act. *Applied Economic Perspectives and Policy, 40*(3), 402-420. https://doi.org/10.1093/aepp/ppx039

Cooley, M., Carychao, D., Crawford-Miksza, L., Jay, M. T., Myers, C., Rose, C., Keys, C., Farrar, J., & Mandrell, R. E. (2007). Incidence and Tracking of Escherichia coli O157:H7 in a Major Produce Production Region in California. *PLoS ONE, 2*(11), e1159. https://doi.org/10.1371/journal.pone.0001159

Hardesty, S. D., & Kusunose, Y. (2009). Growers' compliance costs for the leafy greens marketing agreement and other food safety programs. University of California Small Farm Program Research Brief. https://ucanr.edu/sites/sfp/files/143911.pdf.

Harrison, J. A., Gaskin, J. W., Harrison, M. A., Cannon, J. L., Boyer, R. R., & Zehnder, G. W. (2013). Survey of food safety practices on small to medium-sized farms and in farmers markets. *Journal of Food Protection, 77*(11), 1989-1993. www.doi.org/10.4315/0362-028X.JFP-13-158

Horsfall, S. (2020). *Survey shows detailed traceability systems are in place on LGMA member farms.* California LGMA. https://lgma.ca.gov/news/traceability-systems

Interagency Food Safety Analytics Collaboration (IFSAC) (2018). *Foodborne illness source attribution estimates for 2016 for Salmonella, Escherichia coli O157, Listeria monocytogenes, and Campylobacter using multi-year outbreak surveillance data, United States.* U.S. Department of Health and Human Services, CDC, FDA, USDA-FSIS. https://www.cdc.gov/foodsafety/ifsac/pdf/p19-2016-report-triagency-508.pdf

Marshall, K. E., Hexemer, A., Seelman, S. L., Fatica, M. K., Blessington, T., Hajmeer, M., Kisselburgh, H., Atkinson, R., Hill, K., Sharma, D., Needham, M., Peralta, V., Higa, J., Blickenstaff, K., Williams, I. T., Jhung, M. A., Wise, M., & Gieraltowski, L. (2020). Lessons learned from a decade of investigations of Shiga toxin–producing Escherichia coli outbreaks linked to leafy greens, United States and Canada. *Emerging Infectious Diseases 26*(10), 2319-2328. www.doi.org/10.3201/eid2610.191418

Miller, C. (2019). *Which issues have your attention? [2019 state of the vegetable industry].* Growing Produce. https://www.growingproduce.com/vegetables/which-issues-have-your-attention-2019-state-of-the-vegetable-industry/

Parker, J. S., DeNiro, J., Ivey, M. L., & Doohan, D. (2016). Are small and medium scale produce farms inherent food safety risks? *Journal of Rural Studies, 44,* 250-260.

https://agris.fao.org/agris-search/search.do?recordID=US201600195504

Ribera, L., Palma, M., Paggi, M., Knutson, R., Masabni, J. G., & Anciso, J. (2012). Economic analysis of food safety compliance costs and foodborne illness outbreaks in the United States. *HortTechnology, 22*(2), 150-156. https://doi.org/10.21273/HORTTECH.22.2.150

Rodrigues, C., Oleson, B., & Dunn, L. L. (2020). Food safety needs assessment for Georgia specialty crops. *Journal of Extension, 58*(4), 1-8. https://tigerprints.clemson.edu/cgi/viewcontent.cgi?article=1074&context=joe

Strohbehn, C. H., Enderton, A., Shaw, A. M., Perry, B., Overdiep, J., & Naeve, L. (2018). Determining What Growers Need to Comply with the Food Safety Modernization Act Produce Safety Rule. *Journal of Extension, 56*(7), 1-9. https://tigerprints.clemson.edu/joe/vol56/iss7/14/

U.S. Food and Drug Administration (FDA). (2020a, December). Food Safety Modernization ACT (FSMA). https://www.fda.gov/food/guidance-regulation-food-and-dietary-supplements/food-safety-modernization-act-fsma.

U.S. Food and Drug Administration (FDA). (2020b, September). *Requirements for additional traceability records for certain foods.* Federal Register Vol. 85, No. 1853. https://www.federalregister.gov/documents/2020/09/23/2020-20100/requirements-for-additional-traceability-records-for-certain-foods

U.S. Food and Drug Administration (FDA). (2020c). *Proposed rule: Requirements for additional traceability records for certain foods at-a-glance.* https://www.fda.gov/media/142303/download

U.S. Food and Drug Administration (FDA). (2020d). *New era of smarter food safety blueprint: Modern approaches for modern times.* https://www.fda.gov/food/new-era-smarter-food-safety/new-era-smarter-food-safety-blueprint

U.S. Food and Drug Administration (FDA) (2019). *Outbreak investigation of E. coli: Romaine (November 2018).* https://www.fda.gov/food/outbreaks-foodborne-illness/outbreak-investigation-e-coli-romaine-november-2018

U.S. Department of Agriculture (USDA) (2019, March). *USDA small farm definitions.* Livestock and Poultry Environmental Learning Community. https://lpelc.org/usda-small-farm-definitions/

Ward, A. (2020). *CA LGMA updates food safety practices.* California LGMA. https://lgma.ca.gov/news/ca-lgma-updates-food-safety-practices

JOURNAL OF
Extension

Research in Brief

Volume 60, Issue 2, 2022

Economic Implications of the Mexican Fruit Fly Infestation in Texas

Samuel D. Zapata[1]

AUTHOR: [1]Texas A&M University–College Station.

Abstract. The Texas citrus industry is threatened by the presence of Mexican fruit fly. The objective of this study was to estimate the economic losses caused by this invasive pest. Economic impact is estimated in terms of loss in revenue and increase in operating costs. Under current quarantined areas and pest management strategies, the Texas citrus industry could experience an annual economic loss of $5.79 million. The analysis was extended to evaluate the economic impact associated with different quarantined area scenarios. This article can be used to increase awareness and adapted to estimate the economic impact of emerging invasive pest outbreaks.

INTRODUCTION

The citrus industry is an important agricultural sector in Texas. An estimated 350,000 tons of citrus were produced in Texas during the 2018–2019 growing season, representing a market value of $90 million (U.S. Department of Agriculture (USDA) National Agricultural Statistics Service (NASS), 2019). Most of the citrus production in the state is concentrated in the southern counties of Hidalgo, Cameron, and Willacy.

Due to geographic location, Texas citrus crops are threatened by the constant inflow of Mexican fruit flies (*Anastrepha ludens*). Mexican fruit fly (commonly known as Mexfly) is native to Mexico and Central America and is an invasive pest in South Texas (Texas Department of Agriculture (TDA), 2020). Female flies lay their eggs in a broad range of fruits and vegetables but prefer grapefruit and oranges as citrus hosts (Weens et al., 2015). The damage caused by hatched larvae makes affected fruits unmarketable. However, the main hazard of Mexfly is the risk of introduction to and establishment within agricultural sectors in other regions of the country. To avoid the spread of this invasive pest, the government, in collaboration with the local citrus industry, has implemented eradication protocols within Texas, including quarantine treatments and limited movement of fruits from infested areas (Jang et al., 2015; USDA Animal and Plant Health Inspection Service (APHIS), 2016; Electronic Code of Federal Regulations (e-CFR), 2020). Preventative measures outside the quarantine zone include bait spray applications and Sterile Insect Techniques.

The Mexfly infestation in South Texas is an economic burden to the local citrus industry. Namely, it has resulted in a decrease in revenues and an increase in operating costs. The objective of this article is to assess the direct economic impacts of the Mexfly outbreaks to the Texas citrus industry. This is the first attempt to evaluate the economic implications of actual Mexfly management protocols that permit the interstate movement of citrus fruit from quarantined areas in Texas (USDA APHIS, 2016). The information presented herein can be incorporated into existing Extension programs to increase farmers' awareness about invasive species and their effects on crops (Sundermeier, 2005; Wyatt et al, 2015). Extension professionals have also identified and documented the need for tools and information related to estimating the economic impact of invasive species (Dellinger et al., 2016). To this aim, this article can serve as a reference case and educational instrument for Extension professionals to illustrate in a practical manner the intrinsic risks and potential economic losses caused by invasive pests. Additionally, the proposed valuation method can be adapted to timely analyze the economic impact of other emerging invasive pest and disease outbreaks. Lastly, the findings of this study can serve as supporting data to the assessment and development of cost-effective management strategies for nonindigenous pests.

ERADICATION PROGRAM AND PEST MANAGEMENT

There are strict protocols in place to eradicate Mexfly outbreaks in South Texas and to prevent the establishment of the pest outside of currently-infested areas (Tex. Admin.

Code, 2015; USDA APHIS, 2016; USDA APHIS, 2019). Specifically, there are three citrus system approach options available depending on whether the grove is in a quarantined area under routine preventive Sterile Insect Techniques and the grove's distance from a Mexfly detection (USDA APHIS, 2016). Compliance with the systems approach regulations to mitigate the risks imposed by the Mexfly allows interstate movement of fresh citrus fruits from the entire quarantine area. Given that most commercial citrus operations in Texas are under the USDA preventive release program of sterile flies, Systems Approach Option 1 is considered in this study (Jang et al., 2015).

Systems Approach Option 1 requires regulatory trapping and certified bait sprays to begin 30 days prior to harvest and continue through the end of harvest. Bait spray applications should be conducted weekly during the required period. The two insecticides approved under current eradication protocols are Malathion for conventional groves and Spinosad for organic groves (e-CFR, 2020). However, fruit fly systems approaches can be suspended for the remainder of the season if five or more additional flies or immature stages are detected within a core (USDA APHIS, 2016). When this happens, all fruits from groves partially or totally within the suspended cores are ineligible for interstate movement unless chemically treated, and that fruit is processed for juice. Typically, the citrus harvest season in Texas starts in October and finishes in late April. Monthly preventive bait spray applications are also conducted when eradication protocols are not active and within areas not in quarantine.

BASELINE ECONOMIC IMPACT

This economic analysis focuses only on commercial grapefruit and orange groves, as those two crops represent about 99% of the total citrus acreage in Texas. There are approximately 17,297 acres of grapefruit and 9,024 acres of oranges in the state, including 474 and 150 acres of organic grapefruit and oranges, respectively (Texas Citrus Pest and Disease Management Corporation (TCPDMC), personal communication, March 3, 2020).

At the end of the 2019–2020 harvest season, a total of 1,078.6 mi² were under the Mexfly quarantine in Texas, including 5,635 commercial citrus acres (TDA, 2020; TCPDMC, 2020). Furthermore, 656 commercial citrus acres were located in the cores of the quarantined area (TCPDMC, 2020). For analysis purposes, total citrus acres were proportionally allocated outside and inside of the quarantined area based on actual citrus acreage distribution (Table 1).

This paper adapts the valuation framework proposed by Zapata et al. (2018) to assess the direct economic impacts caused by the Mexfly infestation. The analysis considered a price effect, losses were stratified by citrus variety and control strategy, and the economic impact was estimated using publicly available secondary data. Generally, the overall economic impact of an invasive pest outbreak is measured by the resulting net changes in revenues and operating costs. In the case of the Mexfly outbreak, industry revenues have been affected by a reduction in the sale of fresh fruit from groves within the cores of the quarantined areas. Specifically, the pre-harvest treatments were suspended in 553.75 acres or about 84.41% of the acres within the core because five or more Mexfly larvae were found in 21 commercial groves (TCPDMC, 2020). Fresh fruit from these groves is not allowed for interstate movement; thus, it is sold at reduced citrus juice prices.

The majority of citrus produced in Texas is intended for the fresh fruit market. Based on their quality grading, fruits are marketed as fresh or processed. Small, irregular, and damaged fruits are discounted and sold as processed fruit. Yields and prices used in the analysis for conventional citrus were estimated based on the five-year average yields (i.e., 2014–2019), utilization proportions, and free on board and packinghouse-door prices reported for Texas by USDA NASS (2020). Organic yields were assumed to be 32% and

Table 1. Estimated Distribution of Commercial Citrus Acres in Texas

| | | Out of quarantine | In quarantine | | Total |
			Outside core	Within core	
Grapefruit	Conventional	13,221	3,182	419	16,823
	Organic	373	90	12	474
Oranges	Conventional	6,974	1,679	221	8,874
	Organic	118	28	4	150
Total	Conventional	20,196	4,861	640	25,697
	Organic	490	118	16	624

Note. Based on TCPDMC reported acres as of April 21, 2020. Numbers may not add up due to rounding.

Table 2. Texas Citrus Yields and Prices

		Yield (box[a]/ac)			Price ($/box)	
		Fresh	Processed	Total	Fresh	Processed
Grapefruit	Conventional	325.93	278.07	604.00	16.22	1.52
	Organic	221.63	189.09	410.72	23.10	2.16
Oranges	Conventional	337.23	128.57	465.80	15.31	1.30
	Organic	249.55	95.14	344.69	20.95	1.78

[a]Forty pounds equivalent weight.

Table 3. Mexican Fruit Fly Economic Impact

Economic impact	Preventive	Eradication		Total
		Quarantine excl. core	Core	
Revenue loss	0	0	2,639,143	2,639,143
Grapefruit	0	0	1,741,991	1,741,991
Oranges	0	0	897,153	897,153
Additional operating costs	2,866,623	223,717	57,263	3,147,602
Bait sprays	2,456,181	124,926	44,247	2,625,345
Regulatory trapping	410,441	98,791	13,016	522,248
Total economic impact	**2,866,623**	**223,717**	**2,696,406**	**5,786,746**

26% lower than conventional grapefruit and orange yields, respectively (Savage, 2015). Furthermore, based on the five-year average price differential reported by USDA Agricultural Marketing Service (2020), there were assumed price premiums of 42% for grapefruit and 37% for oranges compared to their conventional counterparts. Table 2 presents the yields and prices used in the analysis.

Under 2019–2020 infestation conditions, the Mexfly outbreak could reduce the overall annual revenue of the Texas citrus industry by $2,639,143 (Table 3). All losses come from the reduced sales of fresh fruits—calculated as the reduction in fresh fruit production multiplied by the fresh fruit price premium—from groves in the core areas affected by the suspension of the fruit fly systems approach. Specifically, the sale of fresh grapefruit could decrease by $1,741,991, while fresh oranges are expected to experience a reduction in sales of $897,153.

On the other hand, operating costs have increased due to the adoption of eradication and preventive protocols to suppress Mexfly populations (Table 3). Namely, added expenditures are given by the cost of eradication and preventive spray applications times the corresponding treated areas. Based on average rates charged by local spraying services, the cost per bait spray application is estimated at $6.00 per acre and $17.50 per acre for conventional and organic citrus groves, respectively. The implementation of mandatory eradication protocols could result in an additional statewide production cost of $169,173. Additionally, preventive insecticide applications have been adopted in about 92% of all commercial citrus acres in Texas (TCPDMC, 2020). Thus, preventive bait sprays cost an estimated additional $2,456,181. The Texas citrus industry also contributes $522,248 annually to support TDA and USDA regulatory Mexfly trapping efforts. For estimation purposes, industry trapping expenses are proportionally allocated based on the current distribution of commercial acres under eradication and preventive regimes. Overall, the total direct economic impact associated with the observed 2019–2020 Mexfly infestation is estimated to be equal to $5,786,746.

ALTERNATIVE QUARANTINE SCENARIOS

Given the rapid evolution of the infestation, it is important to assess the economic implications of potential modification to the Mexfly quarantine areas and their corresponding core areas. To this aim, I conducted a sensitivity analysis to evaluate the economic impact associated with different combinations of total commercial citrus acreage under quarantine, proportion of citrus acres in core areas, and the percentage of core acres with suspended fruit fly system approaches. I evaluated different scenarios based on five potential quarantine zones affecting 0%, 25% (6,580 ac), 50% (13,161 ac), 75% (19,741 ac), and 100% (26,321 ac) of the overall Texas commercial citrus acreage. For each quarantine scenario, I considered three core areas enclosing a total of 5%, 15%, and 25% of the citrus acres within the quarantined area, as well as five potential fruit fly systems approach suspension levels in 25% increments (Table 4). All other parameters and assumptions were equal to the ones described in the previous section. A limitation of the proposed sensibility analysis is its static nature. In practice, the core and quarantine areas, as well as the number of acres with suspended systems approaches, change throughout the growing season.

If Mexfly was eradicated and all related quarantined areas and mandatory eradication protocols were removed in Texas, the citrus industry would still face a total annual economic impact of $1.98 million due to the continuation of preventive actions (Table 4). The total economic burden imposed by the pest increases as more commercial citrus acres are directly affected by the quarantine and by the suspension of the fruit fly systems approaches. For instance, when the systems approaches are active on all citrus acres within the core, no revenue losses are expected and all the

economic impact is generated by the adoption of preventive and eradication protocols. With active systems approaches in place, the total economic impact ranges from $3.81 million to $9.29 million when 25% and 100% of the total citrus acreage is quarantined, respectively. On the other hand, total economic losses substantially increase as the overall number of commercial citrus acres with suspended fruit fly systems approaches rises. For example, overall losses could be equal to $39.68 million (i.e., $31.38 million due to revenue loss and $8.30 million in additional operating costs) if the systems approaches are suspended in 25% of the Texas citrus commercial acreage.

SUMMARY AND CONCLUSIONS

In this article I assessed the direct economic impact caused by the Mexican fruit fly outbreak to the Texas citrus industry. I estimated the economic impact in terms of the loss in revenue and increase in operating costs. A reduction in the value of sales is observed when certified pre-harvest treatments are suspended on citrus groves located within the core of the quarantined areas. Furthermore, additional expenses are generated by the adoption of Mexfly eradication and preventive protocols. I estimated that under 2019–2020 quarantined areas and current pest management strategies, the Texas citrus industry could experience an annual economic impact of $5.79 million, consisting of $2.64 million in rev-

Table 4. Overall Economic Impact of Alternative Infestation Scenarios Evaluated

Proportion of core acres with suspended systems approaches	Quarantined acres						
	0	6,580			13,161		
		Proportion of acres in core					
	0%	5%	15%	25%	5%	15%	25%
	---- million dollars ----						
0%	1.98	3.81	3.81	3.81	5.64	5.64	5.64
25%		4.19	4.95	5.71	6.39	7.91	9.43
50%		4.57	6.09	7.61	7.15	10.19	13.23
75%		4.95	7.23	9.51	7.91	12.47	17.03
100%		5.33	8.37	11.41	8.67	14.75	20.83
		19,741			26,321		
0%		7.46	7.46	7.46	9.29	9.29	9.29
25%		8.60	10.88	13.16	10.81	13.85	16.89
50%		9.74	14.30	18.86	12.33	18.41	24.48
75%		10.88	17.72	24.56	13.85	22.96	32.08
100%		12.02	21.14	30.25	15.37	27.52	39.68

enue loss and \$3.15 million in added operating costs. Also, the magnitude of the economic impact increased as the total citrus acreage affected by the quarantine and the number of core acres with suspended fruit fly systems approaches increased. Particularly, if the Mexfly quarantine is removed, preventive infestation actions could still result in a total annual cost of \$1.98 million. Contrarily, the economic losses associated with more severe infestation situations could be at least \$39.68 million.

The economic impacts presented in this article can be used by Extension professionals to increase awareness about the devastating effects that invasive pests can cause to local economies. The findings highlight the inherent risk associated with the rapid spread of non-native species and the need to develop comprehensive control plans. In this regard, Extension professionals, scientists, industry organizations, and government agencies need to work together to develop and promote cost-effective, research-based pest management strategies. Additionally, the valuation method considered can easily be adapted to promptly assess the economic impact of future pest outbreaks. A timely assessment of the potential risks imposed by novel invasive species is important to implement proper mitigation actions.

REFERENCES

Dellinger, T. A., Day, E. R., and Pfeiffer, D. G. (2016). Brown marmorated stink bug in the Mid-Atlantic states: Assessing grower perceptions, economic impact, and progress. *Journal of Extension 54*(4). https://archives.joe.org/joe/2016august/rb4.php

Domestic Quarantine Notices: Subpart C – Fruit Flies, 7 C.F.R. § 301.32 (2019).

Jang, E., Miller, C., and Caton, B. (2015). Systems approaches for managing the risk of citrus fruit in Texas during a Mexican fruit fly outbreak. U. S. Animal and Plant Health Inspection Service. https://www.aphis.usda.gov/plant_health/plant_pest_info/fruit_flies/downloads/texas-citrus-systems-approach-risk-assesment.pdf

Savage, S. (2015, October). The lower productivity of organic farming: A new analysis and its big implications. *Forbes*. https://www.forbes.com/sites/stevensavage/2015/10/09/the-organic-farming-yield-gap/#639dc3d35e0e

Sundermeier, A. (2005). Exotic pest invasion--Plan of action for Extension educators. *Journal of Extension 43*(5). https://archives.joe.org/joe/2005october/tt5.php

Tex. Admin. Code tit. 4 §19.500 (2015)

U.S. Department of Agriculture. (2020). *Weekly Advertised Fruit & Vegetables Retail Prices*. USDA Agricultural Marketing Service. https://www.marketnews.usda.gov/mnp/fv-report-retail?category=retail&portal=fv&startIndex=1&class=ALL

Federal Order DA-2016–05, 3 C.F.R. (2016). https://www.aphis.usda.gov/plant_health/plant_pest_info/fruit_flies/downloads/spro-fo/DA-2016–05.pdf

U.S.D.A. Animal and Plant Health Inspection Service. (2019). *Fruit Fly Exclusion and Detection: Strategic Plan FY 2019–2023*. https://www.aphis.usda.gov/plant_health/plant_pest_info/fruit_flies/downloads/feed-strategic-plan-en.pdf

U.S.D.A. National Agricultural Statistics Service. (2019, August). *Citrus Fruits 2019 Summary*. ISSN: 1948–9048.

U.S.D.A. National Agricultural Statistics Service. 2020. *Quick Stats*. https://quickstats.nass.usda.gov/

Weens, H.V., Jr., Heppner, J.B., Steck, G. J. (2015). *Mexican fruit fly (Anastrepha ludens)*. Entomology & Nematology at the University of Florida. https://entnemdept.ufl.edu/creatures/fruit/tropical/mexican_fruit_fly.htm

Wyatt, G.J., Herzfeld, D., & Haugen-Brown, T. (2015). Teaching farmers and commercial pesticide applicators about invasive species in pesticide training workshops. *Journal of Extension 53*(5). https://archives.joe.org/joe/2015october/iw7.php

Zapata, S. D., Dudensing, R., Sekula, D., Esparza-díaz, G., & Villanueva, R. (2018). Economic impact of the sugarcane aphid outbreak in South Texas. *Journal of Agricultural and Applied Economics 50*(1): 104–128. https://doi.org/10.1017/aae.2017.24

JOURNAL OF
Extension

Research in Brief

Volume 60, Issue 2, 2022

Fad Diets: Professional Development Needs Among Nutrition Agents in Select Southern States

ABIGAIL P. MCALISTER[1], VICKY L. GREEN[2], SIMONE P. CAMEL[2],
MARY C. FONTENOT[2], AND JANET F. POPE[1]

AUTHORS: [1]Louisiana State University AgCenter. [2]Louisiana Tech University.

Abstract. Because FCS agents teach communities about dietary guidelines, it is important that they are aware of the latest nutrition research. This study assessed FCS agents' knowledge of popular fad diets (low-carbohydrate, intermittent fasting, detoxes and cleanses) and their potential adverse effects, and its relationship among location, certifications, years of experience, professional association memberships, and education. Agents in Louisiana, Texas, Arkansas, and Mississippi were emailed a survey. RDN, DTR, and CHES certifications had significant associations ($p = 0.03$) with knowledge of adverse effects of fad diets. Agents who hold these credentials may have more knowledge about adverse effects of fad diets.

INTRODUCTION

Teaching the public about proper nutrition is an essential component of a Family and Consumer (FCS) agent's role; therefore, agents' awareness of current nutrition and health related trends and what research says about these trends is of great value (Bailey et al, 2014). Fad diets are increasingly popular, and trends frequently change, leaving the public to seek reliable information on the safety and efficacy of these diets (Bailey et al., 2014; Carbone & Zoellner, 2012; Hornick et al., 2013). There is a need to assess the current knowledge of FCS agents pertaining to fad diets in order to develop effective training to support their role (Carbone & Zoellner, 2012).

PURPOSE

The purpose of this cross-sectional study was to assess the general knowledge of FCS cooperative Extension agents concerning fad diets (low carbohydrate, intermittent fasting, detoxes and cleanses) and their potential adverse effects and to determine factors that influence their level of knowledge such as professional certifications held, years of Extension work experience, membership in professional associations, and education level.

METHODS

FCS Extension agents in Louisiana, Texas, Arkansas, and Mississippi were asked to participate in an online survey to identify knowledge gaps and to assess their knowledge, factors influencing their knowledge, and resources used to enhance knowledge. Before data collection began, approval from the Louisiana Tech Institutional Review Board was obtained. A designated person from participating state offices was asked to send the link to the online survey link to agents who were working in FCS, health, nutrition, and related fields. The designated person sending the survey was not requested to report a total number of agents recruited. To ensure confidentiality, the survey was completed in private, on a computer in a location chosen by the participant, and no identifier was assigned to respondents. The survey collected demographic information including geographical state of employment, professional certifications held, years of experience in Extension, membership in professional associations, and education level. The survey's knowledge questions were multiple choice and included the knowledge categories of nutrients provided by foods and adverse effects of fad diets. Content of the knowledge questions was derived from a review of the literature. The survey was reviewed for face validity with three nutrition and dietetics professors and piloted by four adults not employed in CES to check for potential errors in the online format. After this, the survey was revised and finalized.

Agents were given four weeks to complete the survey. State office designees were asked to send a reminder email to these agents in the third week following the initial email. Agents who completed the survey were given the opportunity to enter a drawing for a small gift card; participation in the drawing was voluntary. Identifying information for the drawing was not linked to the survey responses.

Frequency testing was used for descriptive statistical analysis of demographic data. Respondents' knowledge about the premises and potential adverse effects of low-carbohydrate diets (for example, ketogenic and ideal protein diets), intermittent fasting, detoxes, and cleanses was assessed with a numerical test score from survey responses. Knowledge was scored for six categories: knowledge of low-carbohydrate diets, knowledge of intermittent fasting, knowledge of detoxes/cleanses, knowledge of potential adverse effects of low-carbohydrate diets, knowledge of potential adverse effects of intermittent fasting, and knowledge of potential adverse effects of detoxes/cleanses. Scores for knowledge and for potential adverse effects of low-carbohydrate diets, intermittent fasting, detoxes and cleanses were averaged. ANOVA was employed to compare score categories with education level, years of experience, membership in professional associations, certifications held, and the state in which respondents were employed. A $p \leq 0.05$ was considered statistically significant. Statistical analysis was conducted using The Statistical Package for the Social Sciences (SPSS) Version 25 for Students and Stata.

FINDINGS

One hundred eighty-three FCS agents from Louisiana, Texas, Arkansas, and Mississippi responded to the online survey. Response rate was not calculated due to recruitment method. Those who stated they did not practice nutrition education and those who did not complete 90% or more of the survey were excluded from statistical analysis (n=39), resulting in a sample size of 183 respondents for statistical analysis.

Most agents reported general nutrition (93%) and food safety (62%) as their areas of practice. The most frequently reported length of employment was 1–10 years (42%). Education levels varied, but most respondents reported having earned a master's degree (63%) (See Table 1). Over half of the respondents reported Texas as their state of employment (51%) and most were females (96%). A majority of respondents (80%) were members of National Extension Association of Family and Consumer Sciences (NEAFCS). Thirty-one percent indicated they did not hold a nutrition certification. The most frequently reported nutrition certification was Extension Specialist (15%). Of the nutrition certifications, 13 (9%) were Registered Dietitian Nutritionists (RDN) or Nutrition and Dietetics Technician, Registered (NDTR). Information sources most utilized by respondents to obtain information about fad diets included .edu websites (74%), .gov websites (65%), and conference sessions (57%).

Overall, the respondents' mean total score for knowledge (knowledge and adverse effects) was 70%. The average score for knowledge of fad diets was 75%, indicating respondents scored higher on questions regarding knowledge compared to questions regarding adverse effects. The average knowledge score specifically for low-carbohydrate diet questions was 73%, which was the lowest knowledge score among the three fad diets examined. The average knowledge score for intermittent fasting questions was 78%, which was the highest average knowledge score from the three fad diets examined. The average knowledge score for detoxes and cleanses was 74% (Table 2). Potential factors influencing knowledge, including years of work experience in CES, membership in professional associations, education level, and certifications held, served as the independent variables for ANOVA testing. There were no p-values for knowledge scores that indicated significance.

The average knowledge score for adverse effects of fad diets alone was 65%. The average knowledge score specifically for adverse effects of low-carbohydrate diets was 62%, which was the lowest score found for both knowledge and adverse effects. The average knowledge score for adverse effects of intermittent fasting was 64%. The average score for adverse effects of detoxes and cleanses was 69%, which was the highest adverse effects score among the three fad diets examined (Table 2). Potential factors influencing knowledge of adverse effects, including years of work experience in cooperative extension service (CES), membership in professional associations, education level, and certifications held, served as the independent variables. There was a significant main effect for the Certified Health Education Specialist (CHES) certification, $F = 5.16$, $p = 0.03$. There was also a significant main effect for the RDN and NDTR certifications, $F = 4.83$, $p = 0.03$ (Table 3).

CONCLUSIONS, RECOMMENDATIONS, AND IMPLICATIONS

The study results suggest a positive association between CHES certification, RDNs, and NDTRs in the field of cooperative Extension and agents' knowledge of adverse effects of popular fad diets. However, only three respondents had the CHES certification; therefore this association would require further testing. Those who are CHES certified are skilled in all aspects of health education programming, from assessment of community needs to evaluation and serving as a resource for other professionals (Barnes et al., 2002). Thirteen respondents were RDNs and NDTRs., who are trained in dietetics practice and food and nutrition sciences (The Academy Quality Management Committee, 2018). Of the respondents who were RDNs and NDTRs, their years of experience in cooperative extension varied. This may suggest that RDNs or NDTRs, regardless of their years of experience in cooperative extension, have greater knowledge regarding the adverse

Fad Diets: Professional Development Needs Among Nutrition Agents in Select Southern States

Table 1. Respondent Demographics (*n* = 138)

Variable	Respondents *n* (%)
Area of practice	
General Nutrition	128 (93%)
Food Safety	86 (62%)
Weight Control	54 (39%)
Diabetes Management	60 (44%)
EFNEP[a]	20 (15%)
Years of experience	
0–1 year	27 (20%)
>1 year – ≤ 10 years	58 (42%)
> 10 years – ≤ 20 years	31 (22%)
> 20 years	22 (16%)
Education level	
Bachelor's	26 (19%)
Some graduate studies	15 (11%)
Master's	87 (63%)
Doctorate	10 (7%)
State of employment	
Louisiana	15 (11%)
Texas	70 (51%)
Arkansas	36 (26%)
Mississippi	17 (12%)
Gender	
Male	4 (3%)
Female	133 (96%)
Other/I prefer not to disclose	1 (1%)

Note. Respondents may have chosen more than one area of practice. [a]Expanded Food and Nutrition Education Program.

effects of fad diets. Of the thirteen RDNs and NDTRs surveyed, a majority (85%) were members of the Academy of Nutrition and Dietetics.

Agents reported obtaining information from internet-based sources, including .edu websites and .gov websites. While this trend of seeking health information online resembles preferences of the general public, agents in this study often chose more reliable online sources, while the general public tends to choose less reliable sources, like commercial websites (LaValley et al., 2016).

Respondents had a variety of education levels, certifications, years of experience in CES, areas of practice, and professional association memberships, which provided a diverse sample. The convenience of the online survey format allowed participants to respond from work computers. The number of responses varied between states. The time burden (15–20 minutes) may have contributed to some respondents not completing the survey. This sample of agents' responses may not reflect the entire nation's FCS agents; this study could have benefited from a larger sample size. The knowledge questions have not been tested for validity. Finally, while there was a significant association between higher knowledge scores and certain certifications, only thirteen survey respondents were RDNs or NDTRs and only three had the CHES certification, necessitating further research to substantiate this association. Additionally, future research may consider which certifications and professional memberships provide regular continuing education credits and publication subscriptions.

Overall, the survey showed a knowledge deficit among agents regarding fad diets and the adverse effects associated

Table 2. Knowledge Scores Among Professional Certifications Held

Knowledge Subcategory	Extension Specialist	RDN[a] and NDTR[b]	CHES[c]	CHC[d]	CFCS[e]
Low-Carbohydrate Knowledge	72%	83%	67%	86%	67%
Intermittent Fasting Knowledge	77%	80%	83%	71%	78%
Detox and Cleanse Knowledge	75%	76%	84%	78%	77%
Low-Carbohydrate Adverse Effects Knowledge	56%	73%	64%	73%	62%
Intermittent Fasting Adverse Effects Knowledge	63%	65%	71%	61%	64%
Detox and Cleanse Adverse Effects Knowledge	69%	71%	75%	80%	69%
Total Knowledge Score	**69%**	**74%**	**74%**	**74%**	**70%**

Note. Percentages represent the average scores among respondents who hold the professional certifications. [a]Registered Dietitian Nutritionist; [b]Nutrition and Dietetics Technician, Registered; [c]Certified Health Education Specialist; [d]Certified Health Coach; [e]Certified in Family and Consumer Sciences.

Table 3. ANOVA Results for Adverse Effects Knowledge Scores

Variable	F value	p-value
Area of practice		
General Nutrition	0.05	0.83
Food Safety	2.52	0.12
Weight Control	0.16	0.69
Diabetes Management	0.93	0.34
EFNEP[a]	0.22	0.64
Years of experience	0.61	0.61
Membership in professional organizations		
NEAFCS[b]	0.2	0.66
SNEB[c]	0.61	0.44
AND[d]	0.42	0.52
Education level	0.32	0.81
Certifications Held		
Extension Specialist	0.09	0.77
RDN[e] and NDTR[f]	4.83*	0.03
CHES[g]	5.16*	0.03
CHC[h]	0.05	0.82
CFCS[i]	0.31	0.58

[a]Expanded Food and Nutrition Education Program; [b]National Extension Association of Family and Consumer Sciences; [c]Society for Nutrition Education and Behavior; [d]Academy of Nutrition and Dietetics; [e]Registered Dietitian Nutritionist; [f]Nutrition and Dietetics Technician, Registered [g]Certified Health Education Specialist; [h] Certified Health Coach; [i]Certified in Family and Consumer Sciences.
*$p \leq 0.05$.

with them, as the average score among all respondents was 70%. This indicates a need for professional development among FCS agents regarding fad diets. Participating state offices can use the information from this study to develop statewide trainings and resources for their agents.

Over the last decade, there has been a rise in demand for nutrition and health-related programs due to an increased need for improved population health among communities across the nation (Kaufman et al., 2017). CES has evolved to meet the needs of communities by placing a greater focus on nutrition and health programming (Kaufman et al., 2017). As the progression from home economics to nutrition and health in CES continues to develop, there will be a greater demand for nutrition and dietetics professionals, such as RDNs and NDTRs, in CES. In conclusion, a significant association was observed with fad diet adverse effect knowledge scores and RDN, NDTR, and CHES certifications. Results of this study indicate the benefits of hiring CES professionals with these nutrition and health-related certifications and

supporting a variety of continuing education options needed to maintain those credentials or certifications.

REFERENCES

Bailey, N., Hill, A., & Arnold, S. (2014). Information-seeking practices of county Extension agents. *Journal of Extension, 52*(3). https://archives.joe.org/joe/2014june/rb1.php

Barnes, M. D., Neiger, B. L., Mondragon, D., Hanks, W. A., & Brandon, J. E. (2002). Expanded health education roles in managed care: Relationships between CHES, HEDIS®, and NCQA. *Health Promotion Practice, 3*(1), 43–49. https://doi.org/10.1177/152483990200300106

Carbone, E. T. & Zoellner, J. M. (2012). Nutrition and health literacy: A systematic review to inform nutrition research and practice. *Journal of the Academy of Nutrition and Dietetics, 112*(2), 265–265. https://doi.org/10.1016/j.jada.2011.08.042

Hornick, B. A., Childs, N. M., Edge, M. S., Kapsak, W. R., Dooher, C., & White, C. (2013). Is it time to rethink nutrition communications? A 5-year retrospective of Americans' attitudes toward food, nutrition, and health. *Journal of the Academy of Nutrition and Dietetics, 113*(1), 14, 16–23. https://doi.org/10.1016/j.jand.2012.10.009

Kaufman, A., Boren, J., Koukel, S., Ronquillo, F., Davies, C., Nkouaga, C. (2017). Agriculture and health sectors collaborate in addressing population health. *Annals of Family Medicine, 15*(5), 475–480. https://doi.org/10.1370/afm.2087

LaValley, S. A., Kiviniemi, M. T., & Gage-Bouchard, E. A. (2016). Where people look for online health information. *Health Information & Libraries Journal, 34*, 146–155. https://doi.org/10.1111/hir.12143

The Academy Quality Management Committee (2018). Academy of nutrition and dietetics: Revised 2017 scope of practice for the registered dietitian nutritionist. *Journal of the Academy of Nutrition and Dietetics, 118*(1), 141–165. https://doi.org/10.1016/j.jand.2017.10.002

JOURNAL OF
Extension

Research in Brief

Volume 60, Issue 2, 2022

A Needs Assessment Survey of Southern California Pest Management Professionals

SIAVASH TARAVATI[1]

AUTHOR: [1]University of California Cooperative Extension, Los Angeles Office.

Abstract. Pest Management Professionals were surveyed in training workshops/meetings in Southern California between 2015 and 2017 to understand their needs, challenges, and view of integrated pest management. The most encountered pests by pest control technicians were ants, cockroaches, rodents, spiders, termites, and bed bugs. The most challenging pest groups to control were bed bugs, cockroaches, ants, and rodents. The main challenges that professionals faced in doing their work were regulations, followed by managing pests, and customer-related problems. Most participants had a favorable view of IPM. However, they faced several challenges in implementing IPM, among which customer-related issues were the most common.

INTRODUCTION

California, structural pest control licensing is administered by the Structural Pest Control Board (SPCB), a California Department of Consumer Affairs subdivision. California structural pest control technicians, also known as Pest Management Professionals (PMPs), can become licensed in one or more licensing branch: Branch 1, the practice of fumigation using poisonous or lethal gas; Branch 2, the practice of controlling household pests excluding fumigation with poisonous or lethal gases; Branch 3, the practice of controlling wood-destroying pests or organisms by the use of insecticides or structural repairs and corrections, excluding fumigation with poisonous or lethal gases (California Structural Pest Control Board, 2019). As of 23 April 2020, there were 22,719 licensed PMPs in California comprised of 6,086 registered applicators, 12,772 field representatives, and 3,861 operators (California Department of Consumer Affairs, 2020).

In a national survey published by the Environmental Protection Agency (EPA), ants, mosquitoes, and cockroaches were the most common household pests after microorganisms such as bacteria, viruses, mildew, and mold (Whitmore et al., 1992). In another national survey on incidental household nuisance pests, multicolored Asian lady beetles, boxelder bugs, millipedes, and ants were the most common incidental pests of households (Cranshaw, 2011).

Despite the vast expanse of urban areas and the high number of licensed PMPs in California, very few scientific surveys have been performed and published on this vital industry. Limited information is available about California's most common structural pests and the needs and challenges Californian PMPs face in running their pest control businesses. In a survey by Pest Control Operators of California (PCOC), 42.3% of responders chose ants as the biggest problem in California, followed by rats/rodents (31%) and cockroaches (12.6%). When asked, "Which pest is becoming increasingly more difficult to control?" the most common responses were rats/rodents (51.2%), followed by bed bugs (25.6%) (Harbison, 2018).

Integrated Pest Management (IPM) is a pest control approach that utilizes environmentally-friendly methods and tools to reduce pesticide side-effects such as environmental pollution and health risks. Per California state regulations, all PMPs except Branch 1 license holders must undergo IPM training for license renewal, though little is known about the prevalence of IPM use among license holders or the IPM implementation challenges they face. In this paper, I provide results from a new structural pest control survey study on Californian PMPs.

METHODS

I surveyed a total of 182 PMPs in ten in-person pest control training workshops/meetings in Southern California between 2015 and 2017. Participants were provided with an anonymous printed survey form (shown in Figure 1) containing ten questions. Survey takers were allowed to select multiple choices for the categories of "branches" and "counties" and to write down multiple answers for "pests," "knowl-

University *of* California
Agriculture and Natural Resources

UC IPM

Pest Management Survey

By Siavash Taravati (Ph.D) UCCE – Los Angeles 700 W. Main Street - Alhambra, CA 91801

Meeting:	
Company type: ☐Pest Control ☐Product distributor ☐Other (please specify):	
Which branch (es) are you licensed in? ☐Branch I ☐Branch II ☐Branch III ☐n/a	
Which branch (es) are you are actively working in? ☐Branch I ☐Branch II ☐Branch III ☐n/a	
Which counties are you working in? ☐Los Angeles ☐Orange ☐San Diego ☐Imperial ☐Riverside ☐San Bernardino ☐Ventura ☐ Other:	
Question 1: Please name the <u>most common</u> pest(s) that you constantly (or during most of the year) deal with:	
Question 2: Please name the <u>most problematic</u> pest(s) that you deal with (i.e. species which are hard to get rid of in general)	
Question 3: Please name the <u>most common</u> pest(s) species that you usually encounter in <u>each season</u>: Spring: Summer: Fall: Winter:	
Question 4: Do you see a specific major gap in knowledge for controlling urban pests? Please name and explain. (Examples: Lack of good detection devices for drywood termite, lack of understanding the behavior of a particular species, lack of an effective registered chemical for controlling a species)	
Question 5: What are your main challenges or concerns in doing your job? (e.g. post-treatment customer calls, financial concerns, controlling a particular pest species, regulations, etc.)	
Question 6: Do you believe in IPM as an effective tool for reducing environmental and health hazards? ☐yes ☐no	
Question 7: How often do you <u>consider</u> using IPM methods for managing pests:	☐never ☐sometimes ☐often ☐always
Question 8: How often do you <u>use</u> IPM methods for controlling pests?	☐never ☐sometimes ☐often ☐always
Question 9: Do you believe that your business may lose money if it implements IPM techniques? ☐yes ☐no If yes, please explain why and how potentially this conflict can be overcome.	
Question 10: In general, what stops you from implementing IPM in your work? Please name	

Thank you very much for your participation!

Figure 1. Survey form used in this study.

edge gaps," "challenges," and "IPM obstacles." All the forms were collected at the end of each meeting and scanned into PDF files. All the data were manually entered into Microsoft Excel 2013 sheets, where they were processed and analyzed. All the responses were pooled to create a master sheet, and each variable (column) was analyzed separately to produce averages and percentages. For open-ended questions, each answer was manually read and coded into categories. Then,

coded answers were counted and used for producing averages and percentages.

SURVEY RESULTS

Participants held a total of 259 licenses, 9% of which were in Branch 1 (fumigation), 63% in Branch 2 (general household), and 29% in Branch 3 (wood destroying organisms).

A Needs Assessment Survey of Southern California Pest Management Professionals

They worked in the following counties: Los Angeles (28.9%), San Bernardino (19.5%), Riverside (18.8%), Orange (17.7%), Ventura (6.3%), San Diego (5.7%), Imperial (1.3%), Santa Barbara (0.4%), San Luis Obispo (0.4%), and other counties (1.1%). Among the participants, 38% actively worked in one county only, 21% in two counties, 15% in three counties, 18% in four counties, and 9% in five or more counties. Sixty percent of the survey participants were licensed in only one branch, while 34% were licensed in two branches and 6% in three branches. Among pest groups (387 responses), ants and beetles were the most and least common pest groups, respectively (Table 1). Among ants (34 responses), the Argentine ant was the most common (59%) species, followed by fire ants (15%), odorous house ants (12%), rover ants (9%), carpenter ants (3%), and thief ants (3%). Among termites (28 responses), drywood termites (64%) were reported more than subterranean termites (36%). Among cockroaches (30 responses), German cockroach was the most common (53%), followed by American (40%) and Oriental cockroach (7%). Among rodents (34 responses), rats were the most common (71%), followed by mice (24%), gophers (3%), and ground squirrels (3%). Among rats (8 responses), roof rats were more common (75%) than Norway rats (25%).

The most common structural pests in spring were ants (32.6%), termites (17.8%), rodents (13%), cockroaches (12.2%), spiders (10.4%), bed bugs (3.9%), fleas (3%), earwigs (1.7%), mosquitoes (1.3%), crickets (0.9%), flies (0.4%), silverfish (0.4%), carpet beetles (0.4%), bees (0.4%), mold (0.4%), moths (0.4%), and wasps (0.4%). In summer, the most common structural pests included ants (44.2%), cockroaches (17.5%), spiders (11.5%), rodents (5.2%), bed bugs (4.5%), termites (6.7%), fleas (3%), bees (2.2%), flies (1.9%), wasps (1.1%), mosquitoes (0.7%), silverfish (0.4%), earwigs (0.4%), and scorpions (0.4%). Fall structural pests were most commonly ants (24.1%), spiders (12.7%), rodents (20.3%), termites (16%), cockroaches (15.6%), bed bugs (4.2%), crickets (1.9%), earwigs (0.9%), silverfish (0.9%), birds (0.5%), bees (0.5%), wasps (0.5%), scorpions (0.5%), mosquitoes (0.5%), and carpet beetles (0.5%). In winter, the most common structural pests were rodents (56.9%), ants (11.2%), cockroaches (10.6%), termites (6.4%), spiders (5.3%), bed bugs (3.7%), earwigs (1.6%), crickets (1.6%), mosquitoes (1.1%), silverfish (0.5%), and scorpions (0.5%).

The most challenging pest groups to control were bed bugs (36.7%), followed by cockroaches (22.8%). A complete list of pests by their difficulty of control is presented in Table 2.

Forty percent of the participants did recognize one or more knowledge gaps in pest control that need to be addressed, while 60% could not think of any. The knowledge gaps recognized by participants were the lack of efficient products/tools (28%), pest control techniques (25%), inspection or detection methods (21.9%), pest behavior (9%), application methods (6.3%), pesticide resistance (3.1%), identification (3.1%), and IPM (3.1%).

The main challenges that PMPs faced were regulations (27.5%), customer-related problems (16%, including difficulties in customer education, unreasonable customer expectations, lack of customer cooperation, online pesticide shopping by customers, and cheap customers), managing pests (13.5%), post-treatment customer calls (11%), financial problems (10%), hiring/licensing/training technicians (8%), overhead costs (3.5%, including health and business insurance), lack of efficient pest control products (3%), managing time (1%), safety (<1%), insecticide resistance (<1%), and other challenges combined (4%).

Table 1. Most Common Pests Encountered by PMPs in Southern California, USA

Pest	No. of responses	Percentage of total responses
Ants	128	33.1%
Cockroaches	82	21.2%
Rodents	58	15.0%
Spiders	39	10.1%
Termites	36	9.3%
Bed Bugs	36	9.3%
Fleas	4	1.0%
Earwig	2	0.5%
Beetles	2	0.5%
TOTAL	**387**	**100%**

Note. Each participant had the option to write down one or more pest or pest groups on the survey.

Table 2. Most Difficult Pests to Control by PMPs in Southern California, USA

Pest	No. of responses	Percentage of total responses
Common bed bug	79	36.7%
Cockroaches	49	22.8%
Ants	31	14.4%
Rodents	21	9.8%
Termites	14	6.5%
Spiders	13	6.0%
Fleas	4	1.9%
Beetles	2	0.9%
Earwig	1	0.5%
Bees	1	0.5%
TOTAL	**215**	**100%**

Note. Each participant could write down one or more pest or pest group on the survey.

The majority of participants (97.7%) believed that IPM effectively reduces environmental and health hazards, while 2.3% did not believe so. When asked, "how often do you consider using IPM?" the responses were: always (56.5%), often (30%), sometimes (12.5%), and never (1.1%). When asked, "how often do you use IPM?" the responses were always (35.3%), often (34.5%), sometimes (28.3%), and never (1.7%). When asked about the financial impact of implementing IPM, 87.6% of the participants believed that implementing IPM does not negatively affect their business. In comparison, 12.4% of participants believed that implementing IPM may reduce their revenue. The reasons for not using IPM were customer-related issues (62%) such as customer expectations for spraying and lack of customer cooperation, higher amount of time needed for IPM (22%), high cost of IPM (14%), and failure of IPM in controlling target pests (3%).

CONCLUSIONS

To conclude, data reported in this paper provide quantitative and qualitative insight into the Southern California structural pest control industry. These data also provide a baseline for future Extension and marketing efforts on Southern California structural pest control. In a survey conducted in the western United States, participating PMPs involved in bed bug treatment reported customer-related issues such as "lack of preparation", "clutter", "lack of client cooperation", "lack of education", "misinformation", and "high costs" (Sutherland, 2015), which are similar to findings relayed in this article.

In the agriculture sector, a needs assessment survey revealed similar findings in which a lack of a decision-support tool; disbelief in the efficacy of IPM tools; a lack of action/economic thresholds; and a lack of education in pesticide resistance, resistance management, and pest monitoring were some of the major barriers in IPM adoption (Murray et al., 2021).

Ants were the most common structural pest reported in my survey, which is in line with previous reports on national trends in which ants were reported as the most common macroscopic pests in urban areas (Whitmore et al., 1992).

In my survey, bed bugs were reported as the most challenging pest to control in structures which is consistent with similar studies in other US states (Potter et al., 2008). The seasonality of each pest group is an important piece of information that can be useful to Extension agents, pesticide distributors, PMPs, and building managers. For example, results from my survey show that ants were most problematic in or around structures in the summer and least problematic during the winter. In contrast, rodents showed the opposite pattern, being more common around structures in the winter and least common during the summer (Figure 2). Such a difference could be explained by the different physiology of cold-blooded vs. warm-blooded animals. Ants and other arthropods are cold-blooded and are more active in warmer months. Rodents, on the other hand, are warm-blooded and can stay active during fall and winter. Nevertheless, when temperatures drop during the fall, rodents start to seek shelter and move into structures where they become noticed by

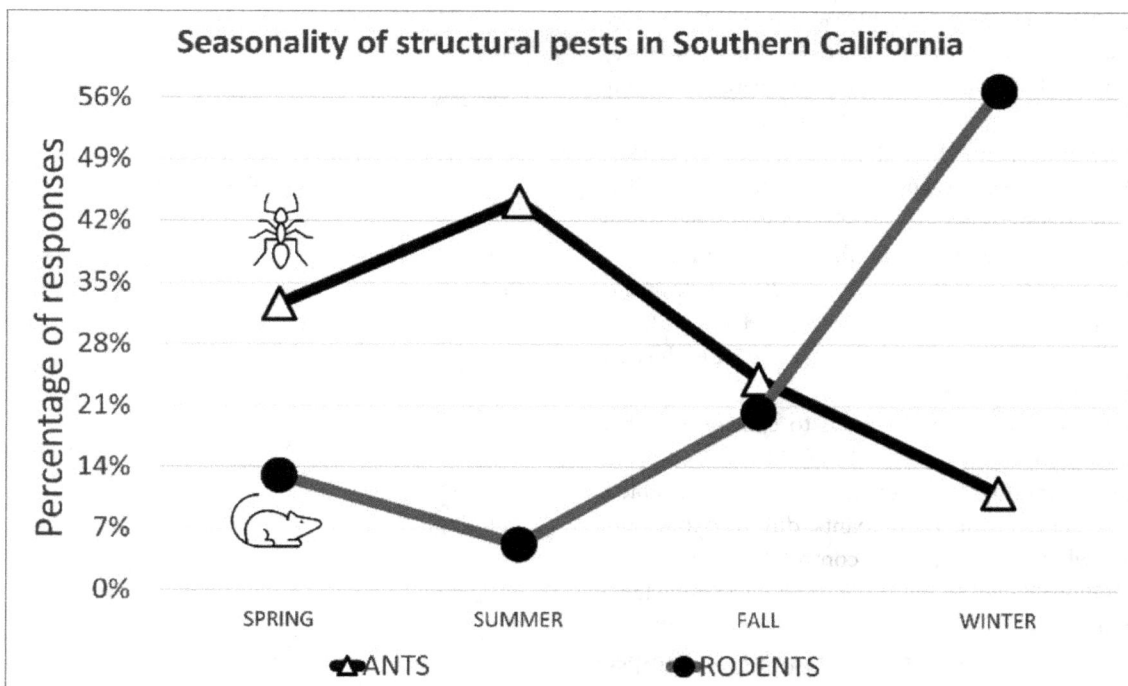

Figure 2. Seasonality of ants vs. rodents' occurrence as structural pests in Southern California.

residents (Frantz & Comings, 1976; Himsworth et al., 2013). While data from this paper show that bed bug activity peaks during the summer, their seasonal variation is much smaller when compared to rodents and ants. By knowing the peak season for each pest, one can efficiently plan to purchase the right pesticides, traps, and application equipment in advance and focus on pest-specific education and marketing before the peak season arrives.

PMPs who participated in the survey had an overall positive view of IPM but had challenges implementing it in their day-to-day work. The most common challenge in implementing IPM was customer-related issues. Therefore, pest control operators and state and local agencies need to educate the public to address this issue before expecting IPM to become more pervasive in structural pest control.

REFERENCES

California Department of Consumer Affairs. (2020). *Public information – Licensee lists overview*. https://www.dca.ca.gov/consumers/public_info/index.shtml

California Structural Pest Control Board. (2019). *How do I become a Structural Pest Control Board licensee?* https://pestboard.ca.gov/howdoi/lic.shtml

Cranshaw, W. (2011). A review of nuisance invader household pests of the United States. *American Entomologist*, *57*(3), 165–169. https://doi.org/10.1093/ae/57.3.165

Frantz, S. C., & Comings, J. P. (1976). Evaluation of urban rodent infestations - An approach in Nepal. *Proceedings of the Vertebrate Pest Conference*, *7*, 279-290. https://escholarship.org/uc/item/8450c5jx

Harbison, B. (2018). *PCOC Survey: Rodent and bed bug infestations on the rise throughout California*. Pest Control Technology. https://www.pctonline.com/article/pcoc-survey-bed-bugs-rodents-surge/

Himsworth, C. G., Feng, A. Y. T., Parsons, K., Kerr, T., & Patrick, D. M. (2013). Using experiential knowledge to understand urban rat ecology: A survey of Canadian pest control professionals. *Urban Ecosystems*, *16*(2), 341–350. doi.org/10.1007/s11252-012-0261-4

Murray, K., Jepson, P., Bouska, C., Scherr, M., & Walenta, D. (2021). Integrated pest management summit reveals barriers, needs, and goals for Agricultural Extension. *Journal of Extension*, *58*(3). https://tigerprints.clemson.edu/joe/vol58/iss3/24/

Potter, M. F., Romero, A., & Haynes, K. F. (2008). Battling bed bugs in the USA. *Proceedings of the Sixth International Conference on Urban Pests*, 401–406.

Sutherland, A. (2015). Bed bug management challenges survey of professional bed bug management in multi-unit housing. *UC IPM's Green Bulletin*, *5*(3), 2–3. https://ucanr.edu/blogs/blogcore/postdetail.cfm?postnum=20177

Whitmore, R. W., Kelly, J. E., & Reading, P. L. (1992). *National home and garden pesticide use survey, final report. Volume I: executive summary, results, and recommendations*. US Environmental Protection Agency.

JOURNAL OF
Extension

Research in Brief

Volume 60, Issue 2, 2022

Facilitators and Barriers to Farmers' Market Use in a Rural Area

Soghra Jarvandi[1], Kristen Johnson[1], and Karen Franck[1]

AUTHORS: [1]University of Tennessee Extension

Abstract. Farmers' markets may improve access to healthful foods in rural areas. Our objective was to identify facilitators and barriers to farmers' market use in a rural county. We collected data via surveys, focus group sessions, and key informant interviews. Study participants identified the two existing farmers' markets as community assets. Barriers to use farmers' markets included inconvenient market hours, not accepting nutrition assistance program benefits, limited transportation, and limited variety. Interventions to improve food access should include ways to meet the needs of specific populations such as low-income residents and residents living in outlying areas without farmers' markets.

INTRODUCTION

Consuming a healthful diet and engaging in physical activity are critical for prevention and management of obesity and chronic conditions. Factors that influence dietary choices include individual, social, physical, and environmental (Story et al., 2008). Examples of environmental influences associated with obesity include areas with limited access to grocery stores (food deserts) and areas with a high presence of fast food and junk food (food swamps) (Cooksey-Stowers et al., 2017).

The use of farmers' markets offers a promising strategy for improving food environments to increase access to healthful foods. Previous studies have shown that farmers' market use is associated with increased intake of fruits and vegetables both among different populations (AbuSabha et al., 2011; Freedman et al., 2013) and in rural communities (Pitts et al., 2014). A systematic review of facilitators and barriers to farmers' market use found several influences on shoppers' decision to utilize markets, including economic, spatial, social, and personal factors as well as factors in service delivery (Freedman et al., 2016). Some consumers view farmers' markets as opportunities to socialize and support local communities and as a source of good quality foods. However, other consumers identify barriers to using markets, such as inconvenient locations and hours of operation (Freedman et al., 2016).

The benefits of farmers' markets extend beyond the consumer to communities and producers (Abel, et al., 1999). Despite these benefits, questions remain about whether farmers' markets are accessible to all individuals (Byker et al., 2013). An analysis of farmers' markets in the United States found that markets are not frequently located within food deserts, but are located in geographic areas with higher socioeconomic status and a higher proportion of white residents (Schupp, 2019). In addition, limited income consumers experience several barriers to use farmers' markets, including not being able to use nutrition assistance program benefits at farmers' markets and limited transportation (Freedman et al., 2016).

With expertise in family and consumer sciences, agriculture and natural resources, and community economic development, Extension is well-poised to engage community members to identify needs and employ interventions that increase accessibility to farmers' markets for all (Civittolo, 2012). Extension's broad reach and community connections within rural and urban areas ensure that existing and newly formed farmers' markets meet the needs of communities, producers, and all community residents (Abel et al., 1999). The objective of the study described in this article was to utilize Extension's expertise and partnerships in a rural Tennessee county to identify facilitators and barriers to farmers' market use among residents.

METHODS, STUDY, POPULATION, AND SETTING

We used data collected in an obesity prevention program, Hardeman Healthy Outreach (H20) for Healthy Weight, led by University of Tennessee Extension in Hardeman County, Tennessee. We refer to this program as "H2O" in short. H20 is a community-based project funded by the Centers for Disease Control and Prevention (CDC) High Obesity Program (HOP). HOP projects target counties with adult obesity rates of 40% or higher using environmental and systems level changes to improve access to healthful foods and increase physical activity.

In 2019, H2O conducted a comprehensive needs assessment to identify community challenges and facilitators for healthful behaviors within each of the nine communities (eight incorporated and one unincorporated towns) in the county, which has a total population of 25,050. For this paper, we included only data related to farmers' market use in the county.

PROCEDURE

The institutional review board for human subjects at the University of Tennessee approved the study protocol. We used several methods to collect data, including surveys, focus groups with county residents, and interviews with key informants. County Extension staff worked to identify, engage, and convene a steering committee consisting of leaders from each community and representatives from the county hospital, the health department, and the school system. Steering committee members helped recruit adults to complete the survey and to participate in focus group sessions. Extension staff identified and recruited professionals for key informant interviews. Focus group sessions lasted for about one hour, and key informant interviews lasted for about 30 minutes.

MEASURES

The survey consisted of questions about access to healthful foods, including a question about whether or not respondents shopped at farmers' markets located in the county. If respondents answered "no," a follow-up question was asked about the reasons for not shopping at farmers markets. For the focus group sessions and key informant interviews, we developed open-ended questions to engage participants and promote discussions about healthful behaviors.

DATA ANALYSIS

We analyzed survey data using descriptive analysis. We audiotaped and transcribed focus group sessions and made detailed notes for the key informant interviews. We analyzed qualitative data through content analysis, using NVivo version 10 to mine the data, create categories, and tag the data (Cresswell, 1998). We used an open coding approach to create and then aggregate the categories (Cresswell, 1998; Krueger, 2014). Two authors (SJ and KF) independently coded the transcripts, then jointly reviewed their findings and discussed any discrepancies to reach consensus.

RESULTS SURVEYS

Table 1 shows the demographics of respondents. Of 1,085 surveys collected, 990 were complete and included in this analysis. Most respondents were female (79%), White (77%), and aged 18–64 years (78%). Demographic compositions in the study sample were different from those of the county (U.S. Census Bureau). Also, 7% of participants reported receiving Supplemental Nutrition Assistant Program (SNAP) benefits, which was less than half of the rate in the county (TN Department of Human Services).

More than half of participants reported that they buy fruits and vegetables at the farmers' markets (Figure 1). Regarding the reasons for not shopping at farmers' mar-

Table 1. Demographics of Survey Participants Compared with the County

	Surveys	Focus Groups	Hardeman County
n	990	61	25,050
Gender, percent			
Female	79%	67%	45%
Male	21%	33%	55%
Age group, percent			
18–64 years	78%	62%	62%
≥ 65 years	22%	38%	19%
Race, percent			
Black	21%	21%	42%
White	77%	77%	55%
Other	2%	2%	3%
Individuals who reported receiving SNAP benefits, percent	7%	–	18%

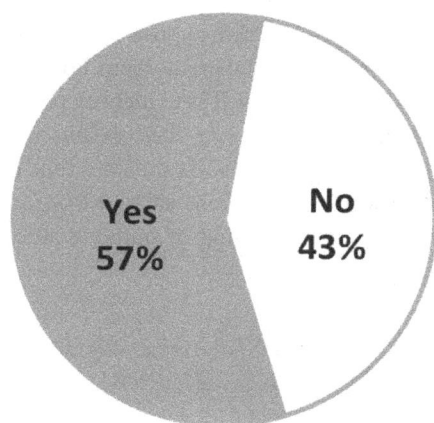

Reported reasons for not shopping at farmers' markets
(frequencies in descending order)

- Market hours of operation 50%
- Not enough variety 14%
- Too expensive 11%
- Other reasons (e.g., inconvenience) 11%
- Not knowing about farmers' market 6%
- Not accepting SNAP 4%
- Poor quality food 3%

Figure 1. Shopping at the farmers' markets.

kets, the most common reported reason was that "they are not open during the hours I usually shop," followed by "not enough variety" (Figure 1).

FOCUS GROUP SESSIONS AND INTERVIEWS

A total of 61 residents participated in nine focus group sessions conducted in different locations across the county. Table 1 shows the demographics of participants. Most participants were female (67%), White (77%), and aged 18–64 years (62%). In addition, a total of 25 professionals and community leaders participated in the key informant interviews, including civic leaders, health care professionals, educators from public schools and Head Start, faith-based leaders, business owners, and volunteers. About half of interview participants were female (48%) and about one fourth were African American (28%).

COMMUNITY ASSETS

This rural county has two farmers' markets located in two different incorporated towns—one in the middle of the county with over 5,000 residents and the other in the southwest corner of the county with almost 500 residents. Participants from six of the nine communities identified the two existing farmers' markets as assets that increased access to healthful foods for residents. Overall, there were 19 positive comments about the farmers' markets. Most of these positive comments (13 of the 19) were made by participants who resided in the two communities where the markets were located.

Positive comments included basic statements about the value of the markets such as: "The farmers' market is good," and "Great farmers' market in [town]." Other positive comments were related to the number of customers: "[Town's] got [a farmers' market] and they do pretty good. I mean they draw a lot of people in."

Participants also commented about the importance of being able to buy fresh produce from local producers. Typical comments included:

> "I try to buy [at the farmers' market] to support the community."

> "I will go to [community name] for the farmer's market because those are our neighbors, so to speak."

BARRIERS

Participants from five of the nine communities identified challenges related to shopping at farmers' markets located in the county. Overall, there were 13 comments related to barriers. Neither of the two farmers' markets accepted SNAP Electronic Benefit Transfer (EBT) and participants identified this as a barrier for low-income customers. Comments included:

> "The new farmers' market is open one day a week . . . but does not accept EBT."

> "Most people here use an EBT card and they can't take that to the farmers' market."

Another barrier was distance to the markets for people who lived outside of the two communities where the markets were located. Comments included: "I am not driving [15 miles away]," and "That's a long way away." However, some commented that they would be willing to drive farther away for markets that had a better variety such as, "[Market outside of the county] is further mileage, but it's just more to get when you're there."

Lack of variety was identified as another barrier. This included markets that focused more on crafts and other goods than food and markets that did not have enough vendors to provide produce. Typical comments included:

> "The farmers' markets here are getting very distracted away from food."

> "The struggle is to get people to bring the food that they do grow that they have left over."

DISCUSSION

The objective of our study was to identify facilitators and barriers to farmers' market usage among residents of a rural county with a high obesity rate. Survey results indicate that over half of respondents shopped at the local markets. Our qualitative findings support the survey results, with several participants identifying the markets as community assets and an important means to support local producers and businesses.

Similar to previous studies (Freedman et al., 2016), we found that farmers' markets are perceived as important sources for healthful foods, and shopping at local farmers' markets is viewed as an opportunity to support local communities. With regards to barriers for not using farmers' markets, survey findings and qualitative findings were different. Survey respondents identified inconvenient market hours as the top barrier. However, participants in focus group sessions and interviews identified additional barriers that are especially relevant for low-income residents, including not being able to use EBT and lack of transportation. Given that only 7% of our survey sample reported participating in SNAP, this may explain why financial and transportation challenges were not highlighted in the survey's findings.

One important aspect to keep in mind is that farmers' markets have to compete with other food retailers and other markets. A struggle in rural farmers' markets can be finding vendors who provide enough produce with a wide variety to entice customers. Our study illustrates that some people are willing to go farther to buy produce from markets in other counties, but residents would prefer to shop nearby to support their farmers and communities

Findings of our study indicate the importance of conducting needs assessments that include multiple methods, when possible, to provide a more comprehensive understanding of community needs. The demographic characteristics of survey respondents and focus group participants differed from the county population in terms of race and gender, and so, our findings may not be representative of all Hardeman County residents. Future studies should include low-income populations to capture their specific challenges related to healthful foods access. Our findings also highlight the importance of identifying community assets and facilitators for healthful behaviors rather than focusing only on

needs and barriers. In initiatives that target health issues like obesity, a strengths-based approach to needs assessment allows community members to focus both on positive aspects of where they live as well as to identify potential solutions to challenges. Furthermore, these results indicate the importance of tailoring food access interventions to address challenges for specific populations such as low-income residents as well as residents living in outlying areas without farmers' markets.

LESSONS LEARNED AND IMPLICATIONS FOR PRACTICE

The results showed that farmers' markets are perceived as community assets. However, barriers may limit usage by certain community members. Extension personnel may benefit from using multiple assessment methods, including surveys, focus group sessions, and key informant interviews, to understand the role of farmers' markets in rural communities. Extension can play an important role in convening multi-sector partnerships to address barriers for farmers' market usage to help ensure accessibility to healthful foods for all community members.

REFERENCES

Abel, J., Thomson, J., & Maretzki, A. (1999). Extension's role with farmers' markets: Working with farmers, consumers, and communities. *Journal of Extension*, *37*(5). https://archives.joe.org/joe/1999october/a4.php

AbuSabha, R., Namjoshi, D., & Klein, A. (2011). Increasing access and affordability of produce improves perceived consumption of vegetables in low-income seniors. *Journal of the American Dietetic Association*, *111*(10), 1549–1555.

Byker, C. J., Misyak, S., Shanks, J., & Serrano, E. L. (2013). Do farmers' markets improve diet of participants using federal nutrition assistance programs? A literature review. *Journal of Extension*, *51*(6). https://archives.joe.org/joe/2013december/a5.php

Civittolo, D. (2012). Extension's role in developing a farmers' market. *Journal of Extension*, *50*(1). https://tigerprints.clemson.edu/joe/vol50/iss1/15/

Cooksey-Stowers, K., Schwartz, M. B., & Brownell, K. D. (2017). Food swamps predict obesity rates better than food deserts in the United States. *International Journal of Environmental Research and Public Health*, *14*(11), 1366.

Cresswell, J. W. (1998). *Qualitative inquiry and research design: Choosing among five traditions*. Sage.

Freedman, D. A., Choi, S. K., Hurley, T., Anadu, E., & Hébert, J. R. (2013). A farmers' market at a federally qualified health center improves fruit and vegetable intake among low-income diabetics. *Preventive Medicine, 56*(5), 288–292.

Freedman, D. A., Vaudrin, N., Schneider, C., Trapl, E., Ohri-Vachaspati, P., Taggart, M., Cascio, M. A., Walsh, C., Flocke, S. (2016). Systematic review of factors influencing farmers' market use overall and among low-income populations. *Journal of the Academy of Nutrition and Dietetics, 116*(7), 1136–1155.

Krueger, R. A. (2014). *Focus groups: A practical guide for applied research.* Sage.

Pitts, S. B. J., Gustafson, A., Wu, Q., Mayo, M. L., Ward, R. K., McGuirt, J. T., Rafferty, A. P., Lancaster, M. F., Evenson, K. R., Keyserling, T. C., Ammerman, A. S. (2014). Farmers' market use is associated with fruit and vegetable consumption in diverse southern rural communities. *Nutrition Journal, 13*(1), 1.

Schupp, J. (2019). Wish you were here? The prevalence of farmers' markets in food deserts: An examination of the United States. *Food, Culture & Society, 22*(1), 111–130.

Story, M., Kaphingst, K. M., Robinson-O'Brien, R., & Glanz, K. (2008). Creating healthy food and eating environments: Policy and environmental approaches. *Annual Review of Public Health, 29*, 253–272.

TN Department of Human Services. (2021). *SNAP-Statistical Information.* TN Department of Human Services. https://www.tn.gov/humanservices/for-families/ supplemental-nutrition-assistance-programsnap/ snap-statistical-information.html

U.S. Census Bureau. (2020). *QuickFacts: Hardeman County, Tennessee.* U.S. Census Bureau. https://www.census.gov/ quickfacts/hardemancountytennessee

JOURNAL OF
Extension

Research in Brief

Volume 60, Issue 2, 2022

Perceptions and Management of Ventenata by Producers in the Inland Pacific Northwest

LISA C. JONES[1], JOHN WALLACE[2], KATHLEEN PAINTER[1], PAMELA PAVEK[3], AND TIMOTHY S. PRATHER[1]

AUTHORS: [1]University of Idaho. [2]Pennsylvania State University. [3]Natural Resource Conservation Service.

Abstract. Ventenata is an annual grass that has invaded agricultural and wildland settings in the Inland Pacific Northwest, causing economic and ecological losses. We know little about producers' perceived risks and management of ventenata. We present results of surveys in 2011 and 2014 targeting producers across affected counties in Idaho and Washington. Awareness of ventenata and costs to producers increased across that time interval. Respondents attending ventenata Extension events adopted recommended management strategies more than those who did not attend. Our study documents the importance of continued integrated pest management research in concert with stakeholder engagement and education.

INTRODUCTION

Researchers of invasion biology have increasingly recognized the importance of human perception when battling invasive species (Kemp et al., 2017; Simberloff et al., 2013). Research development, outreach, and management approaches should consider the influence of stakeholders' perspectives in the awareness, spread, and control of invasive species (Garcia-Llorente et al., 2008).

Ventenata [*Ventenata dubia* (Leers) Coss.] is a nonnative annual grass that has invaded much of the Inland Pacific Northwest (PNW) and northern Rocky Mountains (Alomran et al., 2019). It can displace perennial grasses in Conservation Reserve Program (CRP) land, pastures, rangelands, and hay fields, often causing large reductions in yield and limiting access to export markets (Fountain, 2011; Jones et al., 2018; Wallace & Prather, 2011). Given its ability to successfully invade various agricultural and wildland systems, ventenata's current range is likely smaller than its potential range. Thus, ventenata will continue to present management challenges and costs.

Despite the risk of ventenata, the role of producers in its management remains undeveloped. Specifically, little is known about perception, costs, and management decisions of producers regarding ventenata. Understanding perspectives and practices is critical to development of effective weed management programs (Kelley et al., 2013). To help direct research and Extension efforts, we performed two surveys of producers in Idaho and Washington in 2011 and 2014. The primary objectives were to understand:

- Producers' awareness and risk perceptions of ventenata
- Producers' perceived costs of ventenata infestation
- Producers' management strategies and their perceived efficacy
- If Extension programming on ventenata management techniques were adopted
- If and how producers' perception and management differed across agricultural systems (small grains, grass hay, alfalfa, improved pasture, CRP)

METHODOLOGY

STUDY AREA

The study was conducted in 14 rural counties in eastern Washington and northern Idaho, encompassing 68,096 km² and a population of approximately 875,000 (Figure 1). The region's economy relies heavily upon agricultural production systems, including small grain crops, grass hay, alfalfa, perennial pasture, and CRP land. We chose these counties because they have high ventenata infestation and were focal areas for ventenata-specific Extension efforts.

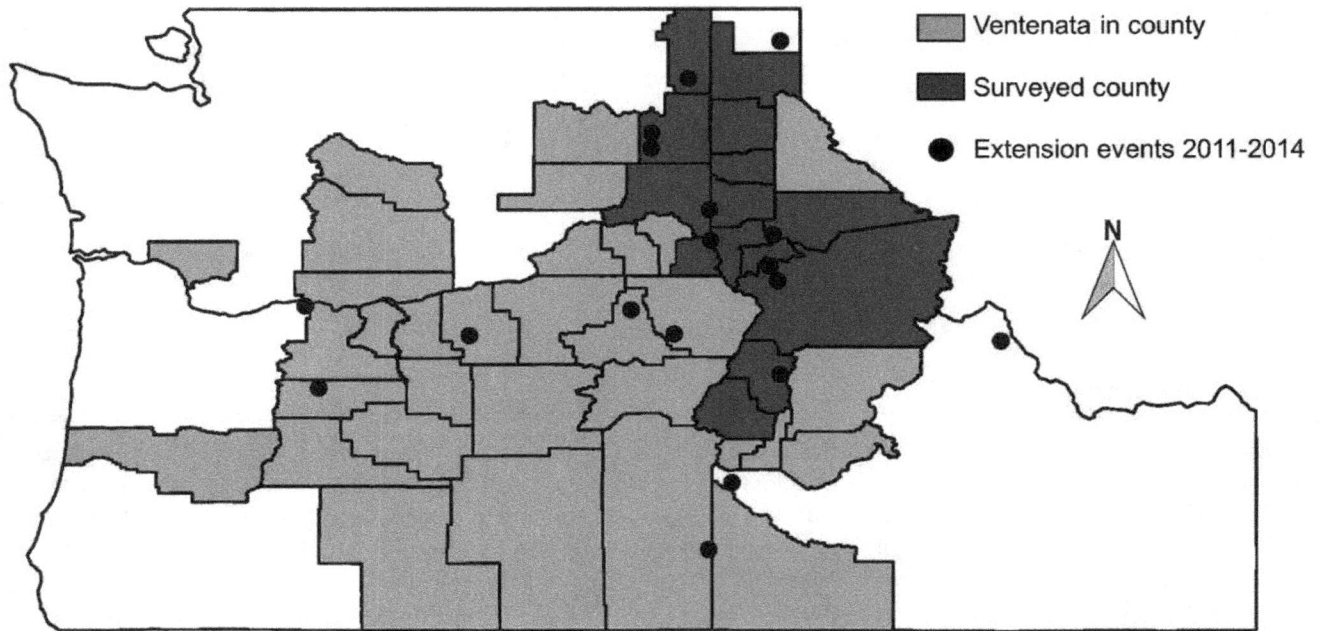

Figure 1. County distribution of ventenata, surveyed counties, and Extension events in Idaho, Oregon, and Washington. Distribution of ventenata (all shaded counties) is based on a survey of natural resources professionals in 2011 (Pavek et al., 2011). Washington counties targeted by the survey were Asotin, Pend Oreille, Spokane, and Whitman. Idaho counties targeted were Adams, Benewah, Bonner, Clearwater, Idaho, Kootenai, Latah, Lewis, Nez Perce, and Washington.

PRODUCER SURVEY

We designed a survey of topics important to both producers and Extension educators that focused on:

- Ventenata awareness and risk perception

- Impacts and costs of ventenata infestation

- Management practices and perceived efficacy

- The role of Extension efforts

The survey was developed and tested with regional producers who collaborated on our Western Sustainable Agriculture Research and Extension grant. Testing the survey involved producers assessing completeness and clarity of the survey with subsequent changes made according to their suggestions prior to distribution. We purchased a random, targeted sample of 1,415 producer addresses in 2011 from Survey Sampling, Inc., and we used the Dillman method for survey contacts (Dillman et al., 2009). We mailed an introductory letter with the survey (see Appendix A) and a photograph of ventenata to producers in March 2011. A reminder postcard was mailed one week later, followed by another copy of the survey one week after that. We obtained a 41.3% response rate (563 completed or partially completed surveys out of 1,362 eligible respondents, with 53 no longer farming).

We used the information provided from the initial survey to guide research and Extension activities over the next three years. Specifically, integrated pest management (IPM)

field studies were performed in CRP and grass hay settings and results were disseminated through 18 Extension workshops or field days throughout the PNW (Wallace et al., 2015; Mackey, 2014).

We sent a follow-up survey in spring 2014 to the respondents who submitted eligible surveys in 2011 using the same mailing sequence. The follow-up survey (Appendix B) included questions from the initial survey and also addressed:

- Attendance at Extension events with information on ventenata

- Consideration and adoption of specific management techniques for controlling ventenata in CRP, grass hay, and pasture

In the follow-up survey we obtained a 54.8% response rate (291 completed or partially completed surveys out of 531 eligible respondents, with an additional 32 no longer farming). The response rates from both surveys met expectations based on the threshold effect of survey length and similar surveys about weeds (Jepson et al., 2005; Johnson et al., 2011).

ANALYSIS

To compare response distributions of categorical data with more than one variable, we used chi-squared tests of contingency tables at the 0.05 significance level to evaluate the hypothesis of no association among variables. In specific instances, interesting outcomes are presented when near the 0.05 significance level and the exact probability is included with the results. To analyze respondents' estimates of infestation and yield reduction, we performed a beta regression with a logit link function. Post hoc tests were performed using linear hypothesis testing. We estimated producer monetary losses from infestation using Idaho prices from the National Agricultural Statistics Service (NASS) and reported yields from survey results.

RESULTS AND DISCUSSION

VENTENATA AWARENESS AND RISK PERCEPTION

Results from the initial survey showed that although fewer than half of respondents had *heard* of ventenata, slightly more than half had seen it (Table 1). More respondents saw ventenata in the follow-up survey (69%) than in the initial survey (56%) ($\chi^2 = 15.99$, $p \leq .01$). Although it is hard to separate the impact of the initial survey on observation, the data were consistent with research suggesting ventenata was undergoing rapid range expansion and becoming more abundant in areas where it previously existed at low densities (Jones et al., 2018; Pavek et al., 2011; Wallace et al., 2015; Anicito, 2013).

From both surveys, a majority of respondents who had heard of ventenata reported they would be very concerned if ventenata established itself on their property (Table 1). Additionally, most respondents considered control to be very or somewhat important (Table 1). There was no difference in either opinion between years (concern $\chi^2 = 1.12$, $p = .57$; control $\chi^2 = 3.1$, $p = .54$).

A summary of producer responses indicated that people heard about and saw ventenata more often in pasture, timothy hay, grass hay, CRP, and non-crop areas, more so than in alfalfa or Kentucky bluegrass. Interestingly, the reported awareness of ventenata was lower in 2014 than in 2011 for each production system (Table 2). Even so, results from both surveys were similar in terms of how ventenata awareness ranked across field types.

IMPACTS AND COSTS OF VENTENATA INFESTATION

In 2011, 54% of respondents reported experiencing increased costs (<$10/ac) due to ventenata and 22% reported larger increased costs (>$10/ac). Most respondents with ventenata reported that ventenata caused them to alter management techniques. Reported costs increased for people's

Table 1. Ventenata Awareness and Perception from the Initial (2011) and Follow-Up (2014) Surveys

Survey question	Response	2011	2014
		% (n)	
Heard of ventenata?*	Yes	47 (262)	73 (209)
	No	53 (295)	27 (76)
Seen ventenata in your county?*	Yes	56 (307)	70 (194)
	No	44 (245)	30 (85)
Level of concern if ventenata established on your property?	Very concerned	61 (191)	57 (156)
	Somewhat concerned	37 (114)	40 (109)
	Not concerned	2 (6)	3 (7)
How important is ventenata control on your property?	Very important	57 (151)	64 (106)
	Somewhat important	35 (91)	29 (48)
	Neither	4 (11)	5 (9)
	Somewhat unimportant	3 (7)	1 (2)
	Very unimportant	1 (2)	1 (1)
Has ventenata increased business costs?*	Yes (>$10/acre)	22 (54)	42 (67)
	Yes (<$10/acre)	54 (131)	32 (50)
	No	24 (58)	26 (41)
Has ventenata altered management?	Yes	69 (170)	74 (129)
	No	31 (76)	26 (42)

Note. Asterisks indicates questions where response patterns differed significantly between 2011 and 2014 (Chi-squared $p \leq .05$).

Table 2. Awareness of Ventenata from the Initial (2011) and Follow-Up (2014) Surveys by Land Use

Field type	Heard it was weedy				Saw it was weedy			
	2011	2014	χ^2	p	2011	2014	χ^2	p
	% (n)				% (n)			
Kentucky bluegrass	54 (70)	35 (45)	11.2	<.01	48 (67)	35 (47)	4.8	.04
Pasture	85 (148)	65 (102)	17.3	<.01	89 (200)	69 (138)	26.5	<.01
Timothy hay	64 (92)	43 (61)	12.3	<.01	64 (115)	45 (67)	12.7	<.01
Grass hay	82 (146)	58 (90)	24.2	<.01	87 (186)	64 (116)	26.5	<.01
CRP	85 (141)	55 (107)	17.8	<.01	84 (177)	69 (130)	13.5	<.01
Alfalfa	52 (71)	32 (47)	11.1	<.01	49 (78)	36 (53)	5.5	.02
Non-crop areas	87 (151)	71 (112)	11.4	<.01	90 (206)	74 (138)	18.8	<.01

Note. CRP refers to the Conservation Reserve Program.

Table 3. Estimates of Ventenata Infestation and Crop Yield Reduction from the Initial (2011) and Follow-Up (2014) Surveys in Different Production Systems

Production system	Average percent estimated infestation (± SE)		Average percent estimated crop yield reduction (± SE)	
	2011	2014	2011	2014
Pasture	25.4[a] (2.2)	25.6[a] (2.7)	30.3[a] (2.8)	27.2[a] (3.0)
Kentucky bluegrass	18.3[a] (4.1)	16.1[a] (3.2)	20.1[ab] (5.5)	17.1[a] (3.8)
Timothy hay	27.9[a] (4.3)	27.3[a] (5.2)	30.1[a] (4.8)	29.1[a] (6.3)
Grass hay	26.8[a] (2.4)	27.8[a] (2.9)	26.0[ab] (2.8)	29.5[a] (3.5)
CRP	21.7[a] (2.3)	20.0[a] (2.5)	29.5[ab] (5.2)	29.1[a] (5.1)
Wheat	18.8[a] (4.7)	—	16.6[b] (5.4)	—
Alfalfa	22.0[a] (4.0)	—	16.7[ab] (3.7)	—

Note. CRP = Conservation Reserve Program. Superscript letters indicate field types that are significantly different in the given year ($p \leq .05$). Dashes indicate data was not obtained.

businesses between 2011 and 2014 ($\chi^2 = 23.608$, $p < .01$), primarily because 47% of people initially reporting a small cost increase subsequently reported a large cost increase. However, there was no change in the percentage of people reporting they were forced to alter their management techniques between 2011 and 2014 ($\chi^2 = 1.99$, $p = .16$). Pre- and post-survey results are compared in Table 1.

To better understand what types of production systems were most heavily affected by ventenata, we asked respondents to report the percent infestation and yield reduction in their fields (Table 3). Estimates of infestation and yield reduction were similar in 2011 and 2014. Higher management costs occurred in pasture, grass hay, and CRP. These systems also had the highest rates of ventenata observations and largest yield reductions. Loss estimates based on Idaho prices and yields of grass hay equaled $89/ac in 2011 and $100/ac in 2014, as calculated using NASS data and our sur-

vey results. For context, on average, 54% of respondents had ventenata in 2014. If those respondents typify Idaho grass hay producers, then as much as 54% of the grass hay (164,000 ac) could become infested, translating to a potential loss of $16 million/year using 2014 NASS data.

MANAGEMENT PRACTICES AND PERCEIVED EFFICACY
The initial survey asked about the diversity and success of management practices producers use to control ventenata. Herbicide application was a common technique and considered highly effective (Table 4). Nearly half of respondents using herbicides used metribuzin or glyphosate, but they used over 25 different herbicides (Figure 2). Because there was no consensus on which herbicides were most effective, educational resources require guidance for herbicide use in combination with other IPM techniques (e.g., Wallace & Prather, 2016).

Perceptions and Management of Ventenata by Producers in the Inland Pacific Northwest

Table 4. Methods of Ventenata Control and Effectiveness of Control Measures in 2011 Survey

Control method	Method used	Method considered most successful
	% (n)	
Crop rotation	26 (59)	10 (19)
Cultivation	46 (112)	18 (35)
Fertilization	16 (43)	4 (9)
Grazing modification	10 (28)	1 (2)
Herbicides	65 (152)	49 (86)
Mowing	60 (149)	13 (20)

In the follow-up survey, respondents were asked whether they adopted or considered adopting specific management practices, all of which were discussed at Extension events (Table 5). Herbicide application was the most common method of ventenata control in timothy or grass hay, CRP, and pasture. Prescribed burning in CRP and pasture was not a common practice, nor was cutting timothy hay to a height of four inches to make timothy more competitive against ventenata. However, IPM research suggests that both methods successfully control ventenata (Mackey, 2014). Future Extension efforts should highlight the efficacy of burning and increasing the cutting height of timothy.

In both surveys, respondents were asked to estimate the level of control they achieved across their production systems. Results indicated that more respondents reported 75% or greater control in 2014 compared to 2011, and fewer respondents reported <50% control after Extension programming (Figure 3). Future research should seek to find control solutions that balance effort, cost, and effectiveness through stakeholder engagement.

THE ROLE OF EXTENSION EFFORTS

Both surveys asked respondents if they accessed Extension or educational materials related to ventenata. More people had accessed such materials in 2014 (28%) compared to 2011 (13%) ($\chi^2 = 31.86$, $p < .01$). In the intervening time, 17% of respondents attended at least one of the 18 Extension events offered in the region. Individuals who attended events were more likely to report experiencing ventenata related losses of more than $10/ac (60%) than those who did not attend (30%) ($\chi^2 = 10.91$, $p < .01$). Producers appeared more engaged in management solutions after experiencing losses due to invasion. This is consistent with other studies suggesting that first-hand experience influences perceptions regarding the importance of invasive species management (Fischer & Charnley, 2012; Johnson et al., 2011). Therefore, Extension efforts should encourage people to adopt preventative measures by increasing awareness of the negative impacts of ventenata.

Attending Extension events increased consideration or adoption of management techniques across production systems (Table 5). In timothy or grass hay, individuals who attended an Extension event were more likely to cut their hay to four inches ($\chi^2 = 3.10$, $p = .08$) or to apply an herbicide ($\chi^2 = 5.23$, $p = .02$) than those who did not attend (Table 5). In combination, these treatments drastically reduce ventenata cover (Mackey, 2014). For those with CRP land, attendees were more likely to burn their fields in the spring ($\chi^2 = 3.18$, $p = .07$) or to consider mowing and spraying in combination ($\chi^2 = 4.41$, $p = .04$) to control ventenata than those who did not attend (Table 5). In pasture systems, attendees were more

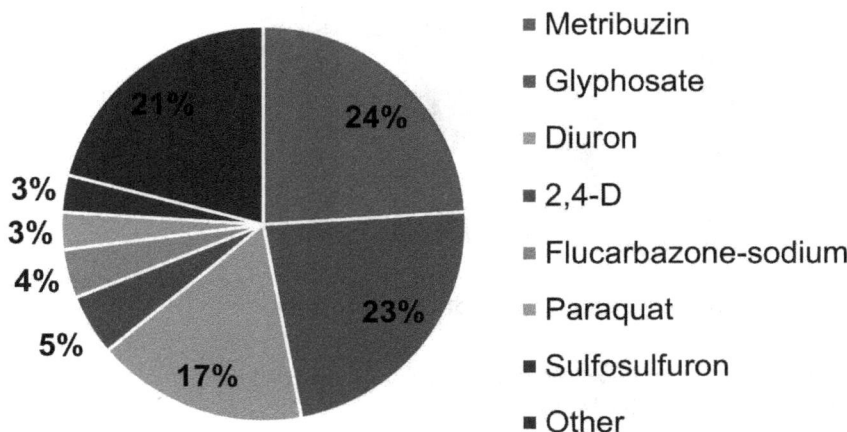

- Metribuzin
- Glyphosate
- Diuron
- 2,4-D
- Flucarbazone-sodium
- Paraquat
- Sulfosulfuron
- Other

Figure 2. Herbicides used for ventenata control by respondents in 2011 survey.

Table 5. Impacts of Attendance at Ventenata-Specific Extension Workshops and Field Days on Management Techniques in Different Field Types

	Adopted technique			Considered technique		
	Attended	Did not attend	Significant	Attended	Did not attend	Significant
Technique	% (n)			% (n)		
Timothy or grass hay						
4 in. harvest height	43 (11)	17 (8)	*	15 (2)	30 (9)	
Fertilize in fall (P, K) and spring (N)	52 (17)	48 (18)		30 (3)	55 (16)	
Apply herbicide	93 (26)	76 (34)	*	91 (10)	84 (26)	
Conservation Reserve Program (CRP)						
Fall burn	11 (2)	11 (3)		37 (7)	20 (5)	
Spring burn	32 (6)	11 (3)	*	35 (6)	16 (4)	
Mow and apply herbicide	71 (17)	60 (21)		91 (10)	56 (10)	*
Apply fertilizer	50 (11)	37 (10)		31 (4)	39 (9)	
Fertilize and apply herbicide	64 (14)	59 (23)		58 (7)	63 (15)	
Pasture						
Fall burn	11 (3)	8 (3)		19 (3)	32 (10)	
Spring burn	12 (3)	14 (5)		29 (5)	33 (10)	
50% livestock utilization	75 (21)	69 (27)		79 (11)	76 (13)	
Fertilize and apply herbicide	79 (23)	58 (22)	*	73 (11)	79 (22)	

Note. Asterisks represent a significant difference in response rates between individuals who attended workshops and field days versus those who did not attend (two-tailed t-test $p \leq .05$).

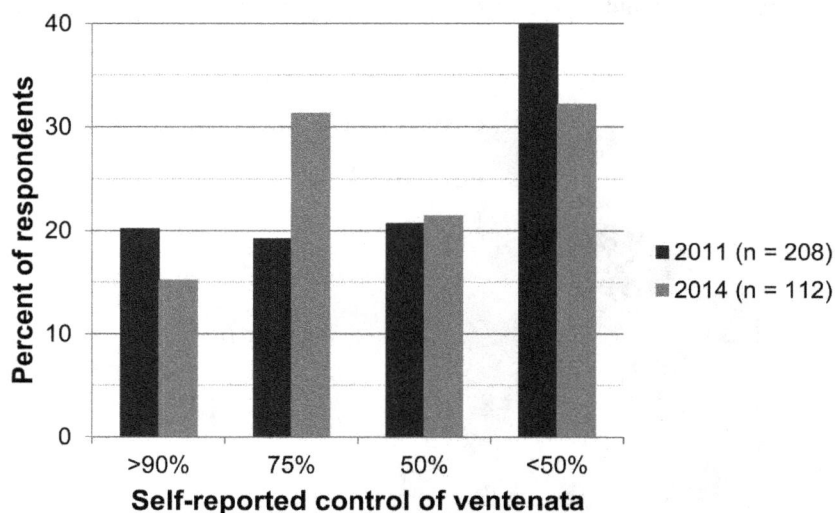

Figure 3. Self-reported levels of ventenata control by respondents from the initial (2011) and follow-up (2014) surveys.

likely to use herbicides and fertilizers in combination (χ^2 = 4.37, p = .04) or herbicides and cattle rotation in combination (χ^2 = 4.22, p = .04) than those who did not attend (Table 5). Our results show that respondents were more likely to adopt or consider suggested management techniques after attending an Extension event, supporting the idea that producers have confidence in Extension education. The confidence corresponds with results from Prokopy et al., (2015) who found that agricultural advisors consider Extension educators to be one of the most trustworthy sources for information. Furthermore, it demonstrated the efficacy of Extension efforts in terms of changing people's attitudes and behaviors (Diem, 2003).

CONCLUSIONS

Our work indicates a high level of awareness and concern about ventenata invasion across a variety of land use types. Additionally, attendees at Extension events were more likely to consider or adopt IPM practices to control ventenata. Stakeholder engagement, through surveys and outreach events, provided opportunities to guide research objectives and disseminate educational materials. Because a proactive approach limits infestation rates and associated control costs, our results imply that Extension educators can reduce costly impacts by targeting stakeholders in vulnerable regions (Wilson et al., 2011).

REFERENCES

Alomran, M., Newcombe, G., & Prather, T. (2019). *Ventenata dubia's* native range and consideration of plant pathogens for biological control. *Invasive Plant Science and Management, 12*(4), 242-245. https://doi.org/10.1017/inp.2019.24

Anicito, K. R. (2013). *A holistic approach to Mima mound prairie restoration* [Master's Thesis, Eastern Washington University]. EWU Campus Repository.

Diem, K. G. (2003). Program development in a political world—It's all about impact! *Journal of Extension 41*(1). http://archives.joe.org/joe/2003february/a6.php

Dillman, D., Smyth, J. D., & Christian, L. M. (2009). *Internet, mail, and mixed-mode surveys: The tailored design method.* John Wiley and Sons, Inc.

Fischer, A. P. & Charnley, S. (2012). Private forest owners and invasive plants: Risk perception and management. *Invasive Plant Science and Management, 5*(3), 375-389. https://doi.org/10.1614/IPSM-D-12-00005.1

Fountain, B. (2011). Producing timothy hay and managing for the impacts of *Ventenata. Proceedings of Western Society of Weed Science*, USA, 107-108. http://www.wsweedscience.org//wp-content/uploads/proceedings-archive/2011.pdf

Garcia-Llorente, M., Martin-Lopez, B., Gonzalez, J. A., Alcorlo, P., & Montes, C. (2008). Social perceptions of the impact and benefits of invasive alien species: Implications for management. *Biological Conservation 141*(12), 2969-2983. https://doi.org/10.1016/j.biocon.2008.09.003

Jepson, C., Asch, D. A., Hershey, J. C., & Ubel, P. A. (2005). In a mailed physician survey, questionnaire length had a threshold effect on response rate. *Journal of Clinical Epidemiology 58*(1), 103-105. https://doi.org/10.1016/j.jclinepi.2004.06.004

Johnson, D. D., Davies, K. W., Schreder, P. T., & Chamberlain, A. (2011). Perceptions of ranchers about medusahead (*Taeniatherum caput-medusae* (L.) Nevski) management on sagebrush steppe rangelands. *Environmental Management, 48*, 400-417. https://doi.org/10.1007/s00267-011-9703-7

Jones, L. C., Norton, N., & Prather, T. S. (2018). Indicators of ventenata (*Ventenata dubia*) invasion in sagebrush steppe rangelands. *Invasive Plant Science and Management, 11*(1), 1-9. https://doi.org/10.1017/inp.2018.7

Kelley, W. K., Fernandez-Gimenez, M. E., & Brown, C. S. (2013). Managing downy brome (*Bromus tectorum*) in the Central Rockies: Land manager perspectives. *Invasive Plant Science and Management, 6*(4), 521-535. https://doi.org/10.1614/IPSM-D-12-00095.1

Kemp, C., van Riper, C. J., BouFajreldin, L., Stewart W. P., Scheunemann, J., van den Born, R. J. G. (2017). Connecting human-nature relationships to environmental behaviors that minimize the spread of aquatic invasive species. *Biological Invasions, 19*(7), 2059-2074. https://doi.org/10.1007/s10530-017-1418-0

Mackey, A. M. (2014). *Developing a decision support tool for ventenata (Ventenata dubia) integrated pest management in the Inland Northwest* [Master's Thesis, University of Idaho]. UI Campus Repository. ProQuest Dissertations Publishing. https://digital.lib.uidaho.edu/digital/collection/etd/id/565

Pavek P., Wallace, J. M., & Prather, T. S. (2011). *Ventenata* biology and distribution in the Pacific Northwest. *Proceedings of Western Society of Weed Science*, USA, 107. http://www.wsweedscience.org//wp-content/uploads/proceedings-archive/2011.pdf

Prokopy, L. S., Carlton, J. S., Arbuckle, J. G., Haigh, T., Lemos, M. C., Mase, A. S., Babin, N., Dunn, M., Andresen, J., Angel, J., Hart, C., & Power, R. (2015). Extension's role in disseminating information about climate change to agricultural stakeholders in the United States. *Climate Change 130*, 261-272. https://doi.org/10.1007/s10584-015-1339-9

Simberloff, D., Martin, J. L., Genovesi, P., Maris, V., Wardle, D. A., Aronson, J., Courchamp, F., Galil, B., Garcia-Berthou, E., Pascal, M., Pysek, P., Sousa,

R., Tabacchi, E., & Vila, M. (2013). Impacts of biological invasions; what's what and the way forward.
Trends in Ecology & Evolution 28(1), 58-66. https://doi.
org/10.1016/j.tree.2012.07.013

Wallace, J. M., Pavek, P. S., Prather, T. S. (2015). Ecological
characteristics of *Ventenata dubia* in the Intermoun-
tain Pacific Northwest. *Invasive Plant Science and
Management 8*(1), 57-71. https://doi.org/10.1614/
IPSM-D-14-00034.1

Wallace, J. M. & Prather, T. S. (2016). Herbicide Control
Strategies for *Ventenata dubia* in the Intermoun-
tain Pacific Northwest. *Invasive Plant Science and
Management 9*(2), 128-137. https://doi.org/10.1614/
IPSM-D-16-00017.1

Wallace, J., & Prather, T. (2011) Management of *Ventenata*
in pasture and CRP. *Proceedings of Western Society of
Weed Science*, USA, 109-110. http://www.wsweed
science.org//wp-content/uploads/proceedings-
archive/2011.pdf

Wilson, J. R. U., Gairifo, C., Gibson, M. R., Ariansutsou, M.,
Bakar, B. B., Baret, S., Celesti-Grapow, L., DiTomaso, J.
M., Dufour-Dror, J. M., Kueffer, C., Kull, C. A., Hoff-
mann, J. H., Impson, F. A. C., Loope, L. L., Marchante,
E., Marchante, H., Moore, J. L., Murphy, D. J., Tassin, J.,
. . . Richardson, D. M. (2011). Risk assessment, erad-
ication, and biological control: Global efforts to limit
Australian acacia invasions. *Diversity and Distributions
17*(5), 1030-1046. https://doi.org/10.1111/j.1472-
4642.2011.00815.x

APPENDIX A. VENTENATA CONTROL PRACTICES SURVEY

The invasive grass, *Ventenata*, which is also known as North Africa Grass is a non-native plant that has established itself in the Pacific Northwest in the past ten years. It primarily affects fields which grow hay and alfalfa, or are used for pasture or CRP. The purpose of this study is to assess the prevalence of the weed in the Inland Northwest, as well as to document the management practices currently in use to control the weed. Your participation in this study will help us improve management practices to control this invasive weed.

Please refer to the color picture included with this survey. The picture is of *Ventenata*, and all of the questions on this survey refer to this plant.

1. Have you heard of the grass, *Ventenata* (sometimes called North Africa Grass), prior to receiving this survey?

     _____   Yes
     _____   No

2. Have you seen the weed *Ventenata* growing anywhere in your county, whether in your fields, on other farms, or along roadsides?

     _____   Yes **→ Go to the next question**
     _____   No **→ Go to Q20, page 7**

3. Where have you seen *Ventenata* growing? (Please mark all that apply)

     _____   Along roadsides
     _____   In pastures
     _____   In rangeland
     _____   Present in CRP acres
     _____   Weedy in CRP acres
     _____   Present in grass hay acres
     _____   Weedy in grass hay acres
     _____   Other _____

4. To the best of your knowledge is *Ventenata* a weed in any of the following crops or situations (either in your fields or the fields of others)? (Please mark all that apply).

Field Type	I have heard of it in these fields (Circle one)	I have seen it in these fields (Circle one)
Wheat	Yes No	Yes No
Pasture	Yes No	Yes No
Kentucky Bluegrass	Yes No	Yes No
Timothy Hay	Yes No	Yes No
Grass Hay	Yes No	Yes No
CRP (Conservation Reserve Program)	Yes No	Yes No
Alfalfa	Yes No	Yes No
Non-crop areas (waste areas/roadsides/equipment yard)	Yes No	Yes No
Grapes	Yes No	Yes No

5. If *Ventenata* became established on your property how concerned would you be?

     _____   Very concerned
     _____   Somewhat concerned
     _____   Not concerned

6. Is *Ventenata* growing on your property?

_____   Yes → **Go to next question**
_____   No → **Go to Q20, Page 7**

7. In what year did *Ventenata* begin growing on your property? _____

8. How important is *Ventenata* control on your property?

_____   Very important
_____   Somewhat important
_____   Neither important nor unimportant
_____   Somewhat unimportant
_____   Very unimportant

9. How much of a problem is *Ventenata* relative to other weeds on your property?

_____   Much more of a problem
_____   Slightly more of a problem
_____   The same level of a problem as other weeds
_____   Slightly less of a problem
_____   Much less of a problem

10. What methods of control have you used on your property?

_____   Cultivation
_____   Mowing
_____   Modifying grazing methods
_____   Herbicides: *please specify type used* _____
_____   Fertilization
_____   Crop rotation: *which crop did you rotate to after growing the affected crop?*

           _____

11. In what year did you begin controlling for *Ventenata?* (If you do not control for *Ventenata,* please leave blank).
_____

12. If you control for *Ventenata* mechanically, how much time do you spend annually on control? _____ hours

13. If you control *Ventenata* chemically, what are the chemical costs per year for control? $_____

14. How frequently do you control for *Ventenata?*

_____   Every year
_____   Every other year
_____   Rarely
_____   I've done it once
_____   Never

15. Which methods have been the MOST successful at reducing *Ventenata,* if any?

_____

_____

_____

16. What percent control have you achieved on your property?

_____ Greater than 90% control
_____ About 75% control
_____ About 50% control
_____ Less than 50% control

17. Which crops has *Ventenata* affected on your property? (Please enter 0% if it does not appear in that field type)

Crop/Field Type	Estimated Percent Infestation in Fields	Estimated Percent Reduction in Yield
Wheat	%	%
Pasture	%	%
Kentucky Bluegrass	%	%
Timothy Hay		
Grass hay	%	%
Conservation Reserve Program (CRP)	%	%
Alfalfa	%	%
Other _____	%	%
Other _____	%	%

18. Has *Ventenata* increased costs to your business in the affected crops?

_____ No
_____ Yes, a little bit
_____ Yes, a great deal

19. Has *Ventenata* altered how you manage your operation?

_____ No
_____ Yes, a little bit
_____ Yes, a great deal

20. Is *Ventenata* control required by FSA in your county?

_____ No
_____ Yes
_____ I don't have CRP

21. Has *Ventenata* in pasture caused you to take any of the following actions?

_____ I don't have *Ventenata* in pasture
_____ Alter stocking rates
_____ Change rotations
_____ Other _____

22. Have you accessed any Extension related products or educational materials related to *Ventenata*?

_____ No
_____ Yes

23. How would you prefer to receive information on *Ventenata* management from Extension?

　　_____　　Publications
　　_____　　Website
　　_____　　Field days/demonstrations
　　_____　　Other_____

24. What type of crops/livestock do you currently have on your operation?

　　_____　　Wheat
　　_____　　Corn
　　_____　　Pasture for cattle
　　_____　　Pasture for sheep
　　_____　　Pasture for horses
　　_____　　Kentucky Bluegrass seed
　　_____　　Alfalfa
　　_____　　CRP
　　_____　　Peas
　　_____　　Grass hay
　　_____　　Other _____

25. How many acres do you have under production (whether owned or leased)? _____ acres

26. How many years have you been involved in agricultural production? _____ years

Do you have any additional comments you'd like to share?

Thank you for your time. Please return this questionnaire in the self-addressed, stamped envelope provided.

APPENDIX B. VENTENATA CONTROL PRACTICES FOLLOW-UP SURVEY

The invasive grass, *Ventenata*, which is also known as North Africa Grass is a non-native plant that has established itself in the Pacific Northwest in the past twelve years. It primarily affects fields which grow grass hay, or are used for pasture or CRP. The purpose of this follow-up study is to assess if education measures helped producers improve management practices in controlling this invasive weed.

1. Have you heard of the grass, *Ventenata* (sometimes called North Africa Grass), prior to receiving this survey?

_____ Yes
_____ No

2. Have you seen the weed *Ventenata* growing anywhere in your county, whether in your fields, on other farms, or along roadsides?

_____ Yes
_____ No

3. To the best of your knowledge is *Ventenata* a weed in any of the following crops or situations (either in your fields or the fields of others)? (Please mark all that apply).

Field Type	I have heard of it in these fields (Circle one)	I have seen it in these fields (Circle one)
Rangeland	Yes No	Yes No
Pasture	Yes No	Yes No
Kentucky Bluegrass	Yes No	Yes No
Timothy Hay	Yes No	Yes No
Grass hay	Yes No	Yes No
CRP (Conservation Reserve Program)	Yes No	Yes No
Alfalfa	Yes No	Yes No
Non-crop areas (waste areas/ roadsides/ equipment yard)	Yes No	Yes No
Other_____	Yes No	Yes No
Other_____	Yes No	Yes No

4. If *Ventenata* became established on your property how concerned would you be?

_____ Very concerned
_____ Somewhat concerned
_____ Not concerned

5. Is *Ventenata* growing on your property?

_____ Yes → **Go to next question**
_____ No → **Go to Q20, page 7**

6. How important is *Ventenata* control on your property?

_____ Very important
_____ Somewhat important
_____ Neither important nor unimportant
_____ Somewhat unimportant
_____ Very unimportant

7. What percent control have you achieved on your property?

_____   Greater than 90% control

_____   About 75% control

_____   About 50% control

_____   Less than 50% control

8. Which crops has *Ventenata* affected on your property? (Please enter 0% if it does not appear in that field type).

Crop/Field Type	Estimated Percent Infestation in Fields	Estimated Percent Reduction in Yield
Pasture	%	%
Kentucky Bluegrass	%	%
Timothy Hay	%	%
Grass hay	%	%
Conservation Reserve Program (CRP)	%	%

9. Has *Ventenata* increased costs to your business in the affected crops?

_____   No

_____   Yes, < $10/acre

_____   Yes, >$10/acre

10. Has *Ventenata* altered how you manage your operation?

_____   No

_____   Yes, a little bit

11. Do you grow grass hay?

_____   Yes

_____   No → **Skip to Q14**

12. With respect to the following management practices in timothy or other grass hay, circle your answer:

	Adopted the technique (Circle one)	Considering adoption (Circle one)
Harvest timothy hay at 4 inch height to make timothy more competitive	Yes No	Yes No
Apply potassium and phosphorus in the fall and nitrogen in the spring to make timothy competitive	Yes No	Yes No
Apply an effective herbicide when one is registered	Yes No	Yes No

13. Does *Ventenata* contamination affect export of your grass hay?

_____   Yes

_____   No

_____   I don't export my hay

14. Do you manage CRP?

_____ Yes

_____ No → **Skip to Q17**

15. With respect to the following mid-contract management practices in CRP, circle your answer:

	Adopted the technique (Circle one)	Considering adoption (Circle one)
Burn in fall to control *Ventenata*	Yes No	Yes No
Burn in the spring to control *Ventenata*	Yes No	Yes No
Mow to rejuvenate stand and spray an herbicide for *Ventenata* control	Yes No	Yes No
Apply fertilizer if soil is deficient to make grasses more competitive	Yes No	Yes No
Spray an herbicide and fertilize for *Ventenata* control and make grasses competitive	Yes No	Yes No

16. Is *Ventenata* control required by Farm Service Agency (FSA) in your county?

_____ No

_____ Yes

17. Do you manage pasture?

_____ Yes

_____ No → **Skip to Q20**

18. With respect to the following management practices in pasture, circle your answer:

	Adopted the technique (Circle one)	Considering adoption (Circle one)
Burn in fall to control Ventenata	Yes No	Yes No
Burn in the spring to control Ventenata	Yes No	Yes No
Rotate cattle to a different pasture when 50% of the forage has been removed	Yes No	Yes No
Spray an herbicide and fertilize for Ventenata control and to make grasses competitive	Yes No	Yes No
Spray an herbicide along with livestock rotation when 50% of forage has been eaten	Yes No	Yes No

19. Has *Ventenata* in pasture caused you to take any of the following actions?

_____ Alter stocking rates

_____ Change rotations

_____ Other _____

_____

_____

20. Have you accessed any Extension related products or educational materials related to *Ventenata*?

    _____   No

    _____   Yes

21. Have you attended a workshop or field day that discussed *Ventenata*?

    _____   No

    _____   Yes, once

    _____   Yes, twice

    _____   Yes, three or more time

22. How would you prefer to receive information on *Ventenata* management from Extension?

    _____   Publications

    _____   Website

    _____   Field days/demonstrations

    _____   Other _____

                                        _____

23. How many acres do you have under production (whether owned or leased)?

    _____ total acres

24. How many years have you been involved in agricultural production? _____ years

Do you have any additional comments about *Ventenata* you'd like to share?

Thank you for your time. Please return this questionnaire in the self-addressed, stamped envelope provided.

JOURNAL OF Extension

Research in Brief
Volume 60, Issue 2, 2022

Michigan Conifer Growers' Perspectives on Disease Management

EMILY S. HUFF[1] AND MONIQUE L. SAKALIDIS[1]

AUTHORS: [1]Michigan State University.

Abstract. A survey of commercial nursery and Christmas tree growers was implemented online and by mail in 2018 to understand disease issues and information preferences. Overall, the majority of the respondents reported that they prefer online sources of information and many are using Extension bulletins and sources. Cultural, chemical, and weed control methods were considered extremely effective by participants and very few used biological control methods to control disease. Participants identified spruce decline, boxwood blight, and hemlock woolly adelgid as emerging disease threats, so future information to growers should focus on identification and management of these threats.

INTRODUCTION

The nursery crop industry in the United States was valued at $4.65 billion as of 2007 (National Agricultural Statistics Service, United States Department of Agriculture [NASS, USDA], 2007b). The economic impact of Michigan's nursery, perennial production, Christmas tree, turf production, landscaping, and lawn care industries was $5.71 billion as of 2012 (Knudson & Peterson, 2012). Climate, soils, and a centralized location to large domestic markets combine to make Michigan a national leader in the production of landscape nursery stock and Christmas trees. Michigan is the third largest producer of Christmas trees in the United States behind Oregon and North Carolina. Michigan ranks 11th in the nation in nursery stock sold. The most recently available report, published in 2008, indicated that Michigan ranked 10th nationally in nursery worker employment (with 7,555 permanent and temporary jobs) and approximately 57% of total production was sold wholesale with 31% being sold wholesale to landscape service firms (Hall et al., 2020). Diseases are important limiting factors in crop production and are also drivers of increasing operation costs; there were 3.9 million pounds of chemicals applied to nursery crops in 2006 (NASS, USDA, 2007a). Disease issues affecting conifers in Michigan include a range of fungal and oomycete pathogens that attack the roots, wood, and foliage of these plants throughout different age classes. These pathogens cause needle loss, branch and tip dieback, and, in some cases, tree and seedling mortality. Some key pathogens the affect conifers include the suite of fungal pathogens behind "Spruce Decline," unculturable rust diseases such as Weir's cushion rust, and the root rot *Phytophthora* species.

Extension educators; nursery, seedling, and Christmas tree growers; and other industry professionals in Michigan have identified the need for efficient and rapid diagnoses of diseases affecting conifers and other woody ornamentals as a pressing issue. These concerns, however, have been expressed by specific individuals; the extent to which the entire conifer grower community in Michigan faces these challenges is unknown. Crop-growing conditions (monoculture) and the addition of moisture and nutrients promote pathogen proliferation, so early detection, efficient containment, and eradication of plant pathogens is crucial. Once a plant pathogen is suspected, it must be properly identified and surveyed. Alongside timely and accurate identification of disease-causing agents, mitigation strategies need to be updated or developed. Current chemical treatment plans are based on anecdotal grower experiences or recommendations from university educators and other industry professionals. The labels on treatment chemicals give instructions limited to few species and may not be legally used in tree species that are commonly cultivated. There is an urgent need to review currently used disease treatments and scientifically validate them. Similarly, a new methodology that objectively identifies the most at-risk tree crops and the most serious pathogens and mitigation strategies of concern to growers, diagnostic labs, agencies, and extension educators is needed. Finally, research findings need to be communicated back to stakeholders and end users using optimized information, communication materials, and methods.

Therefore, there is a strong need to understand the challenges and issues faced by conifer growers to provide timely and effective Extension services to these individuals and to support their efforts statewide. This study implemented a statewide survey to systematically evaluate stakeholders and end user's needs, concerns, and perspectives in relation to conifer disease issues.

METHODS

A hybrid survey was distributed through mail and online in 2018 to a list of all known Michigan nursery, seedling, and Christmas tree growers, as provided by Extension educators working in the state (N = 587). The survey instrument was developed with a group of key stakeholders and included the following sections: Background, Species and Diseases, Effectiveness of Current Practices, Barriers, Information Seeking Behaviors, Monitoring Behaviors, Regulatory Pressures, and Research Needs. The survey was first sent to LISTSERVs maintained by Extension educators via an online link generated in the Qualtrics survey platform. Then, postcards, mail surveys, and follow-up surveys for non-responders were prepared and sent to firms across the state. The results reported below may not add to 100%, as participants could select more than one option. Additionally, there were individuals that did not respond to certain questions (item non-response); percentages reflect only a portion of the number of individuals in a given category that responded to the question. This is a more conservative summary of results, as it is impossible to speculate why an item was left blank.

RESULTS

There were 30 mail surveys and 16 emails that were returned as undeliverable. A total of 87 responses were received via mail and by the online link of the survey for an adjusted response rate of 16%. A statistical comparison of early and late responders to test for non-response bias (ANOVA, R statistical software) revealed no significant differences in the type of farm, number of trees managed, or a composite number of conifer species grown. Responses were received from 36 counties and the average number of acres cultivated by respondents was 356.

Of the respondents that answered what type of farm they operated (n = 80), 47% grew Christmas trees, 31% grew large Ball and Burlap nursery trees, and 17% grew container nursery trees. Less than 10% grew bare root seedlings, potted liners, and plastic liners in greenhouses.

Half of respondents grew some type of coniferous tree: 24% grew deciduous trees, 16% grew shrubs, 14% grew woody perennials, and 8% grew ground cover. Results do not add to 100%, as some respondents grew multiple tree types. The most commonly grown species were Colorado blue spruce (*Picea pungens*), white pine (*Pinus strobus*), Fraser fir (*Abies fraseri*), Norway spruce (*Picea abies*), Black Hills spruce (*Picea glauca* var. densata), concolor fir (*Abies concolor*), and Canaan fir (*Abies balsamea* var. phanerolepis). The most common diseases encountered were Rhizosphaera needle cast (called needle blight on the survey) of spruce, Phytophthora root rot, spruce decline, Diplodia tip blight, and Rhabdocline needle cast, as measured by the number of times a respondent listed the disease in the Top 10 Diseases Encountered question. Respondents most commonly avoided planting Colorado blue spruce and Douglas fir due to concerns about disease vulnerability.

Respondents indicated that cultural (e.g., water mechanical and site selection), weed control, and chemical control measures were quite effective (Figure 1), while biological agents were not typically used by growers. The most commonly reported barrier to disease control was lack of effective products (10%) followed by lack of information (8%).

The most commonly used sources of information by respondents were Michigan State University (MSU) Extension bulletins and websites (n = 18) followed by the MSU Christmas Tree pest management guide (n = 16). Fifteen respondents personally contacted MSU Extension educators, and 14 respondents reported contacting other growers. Only 11 respondents indicated that they needed additional materials, which included resources for identifying diseases and timing pesticide applications. The emerging diseases indicated by respondents were spruce decline, boxwood blight, and hemlock woolly adelgid. Nearly half of respondents (n = 35) reported having monitoring/scouting programs, with the most common timing being whenever they are out in the field. Finally, respondents' preferred modes of communication were Extension bulletins (n = 31) and websites (n = 28). Of the 32 respondents who answered the question about participating in a future annual growers survey, 81% were willing to participate. Of those willing to participate, 90% preferred to answer a questionnaire in the winter and 95% preferred a web-based format. The remaining 5% preferred either a paper questionnaire or an in-person format.

DISCUSSION

With fewer extension educators covering expanding industries and the limited time of growers to attend in person meetings, or even listen to webinar recordings, methods of efficient engagement are critical to ensure stakeholders and the extension community receive timely and relevant knowledge. Methods like annual online surveys may ensure that resources are directed towards critical research to serve these communities, as defined by the individual growers in those communities.

A 30-question survey was distributed to 587 conifer growers in Michigan; 16% of the surveys were returned

Control measure effectiveness

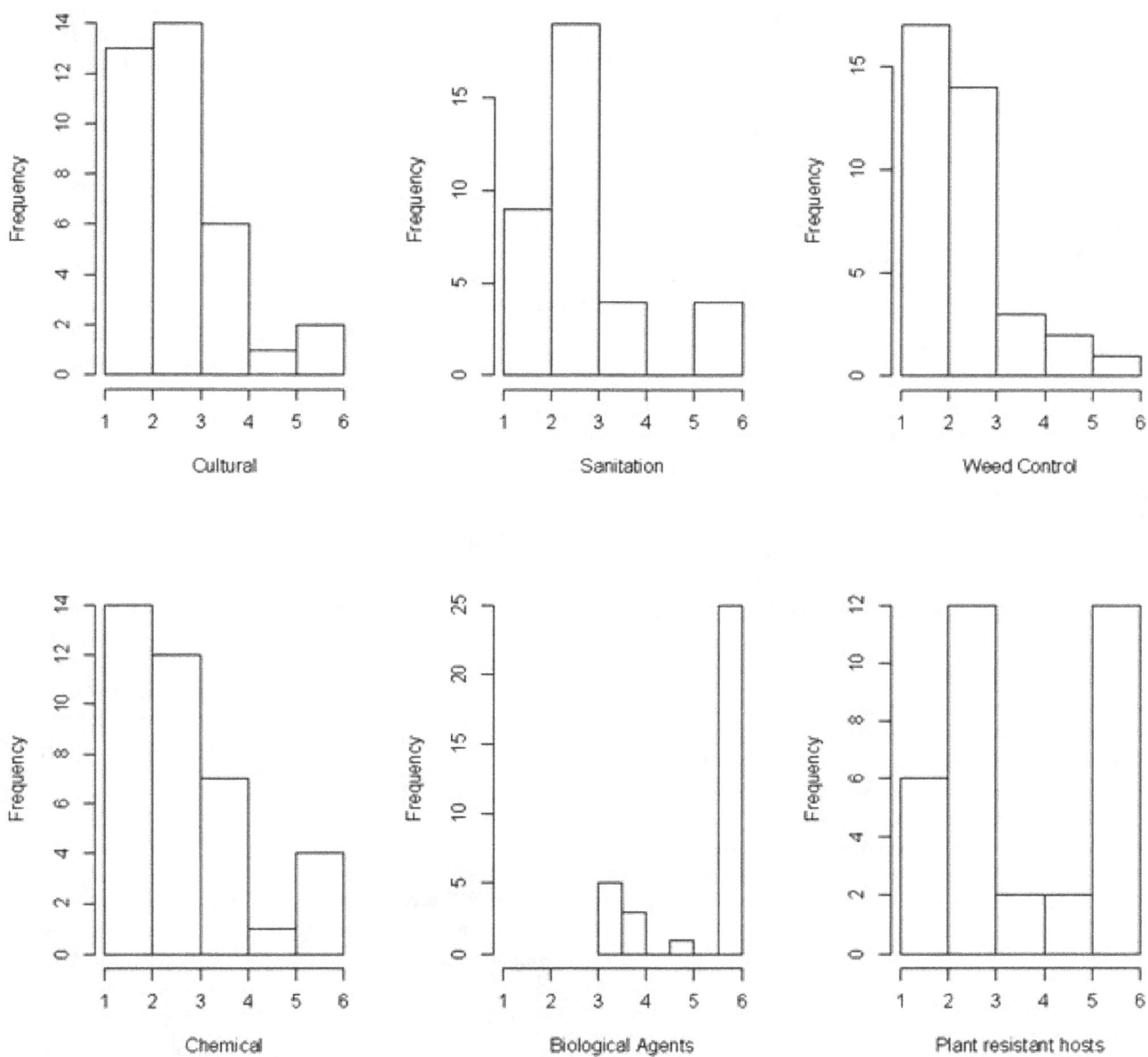

Figure 1. Participant ranking of control measure effectiveness. 1 = *Extremely effective*, 5 = *Not effective at all*, 6 = *I do not use this control measure*.

(n = 87 respondents), and respondents represented both the lower and upper peninsula of Michigan. While respondents indicated a variety of preferences for receiving information, which is consistent with similar efforts to understand grower needs (e.g., Alston & Redding, 1998), most participants preferred online formats. Online communication could take many forms, including traditional Extension outlets and social media (Darr et al., 2020).

Allowing growers to skip questions did mean that all requested information was not captured for each respondent. However, this option was included to enable more respondents to fill in the survey overall. Using multiple choice for some questions and including a text option for others—where respondents could type in their own answer—was important for capturing information that may not have been gathered using preset answers. The online version of the survey was preferred, but there were still some growers that preferred to respond (and did) by mail, so a hybrid approach seems appropriate for the foreseeable future.

Responses indicated that while Colorado blue spruce and Fraser fir accounted for the most commonly grown tree species, respondents avoided planting these trees due to disease concerns. Needle casts and blights (Rhizosphaera needle cast of spruce, Diplodia, and Rhabdocline needle cast), Phytophthora root rot, and spruce decline were considered the most common diseases encountered. Respondents indicated a lack of chemical control options limiting their management of these diseases. Considering Colorado blue spruces are affected by Rhizosphaera and spruce decline and Fraser firs are affected by Phytophthora root rot, effective control of these diseases is critical for industry expansion and stability. Grower perceptions of effective control methods indicated widespread support for cultural, chemical, and weed control methods but little support for biological agents. Therefore, more knowledge and skill building around biological agents will be necessary with this audience.

Future research efforts similar to this would ideally tailor questions to the issues facing growers and would facilitate fast and efficient information gathering (e.g., easy check boxes with lists of diseases). It may be prudent not to allow participants to skip questions (in the online version) so that better data is captured; including a "does not apply to me" option would be critical in this case. A future survey should also include images of typical signs and symptoms of a disease to ensure respondents are selecting the disease even if they do not know the name. Basic pest and pathogen identification has been identified as a knowledge gap in other survey efforts (Byamukama et al., 2016). Assessing growers' general knowledge of diseases and emerging threats would be a useful addition to this effort, as would further exploration of growers' best management practices (Fain et al., 2000).

REFERENCES

Alston, D. G., & Reding, M. E. (1998). Factors influencing adoption and educational outreach of integrated pest management. *Journal of Extension, 36*(3) https://archives.joe.org/joe/1998june/a3.php

Byamukama, E., Szczepaniec, A., Strunk, C., Fanning, R., Bachmann, A., Deneke D., & Johnson P. (2016). Assessing integrated pest management implementation and knowledge gaps in South Dakota. *Journal of Extension, 54*(1). https://tigerprints.clemson.edu/joe/vol54/iss1/11/

Darr, M., Hulcr, J., Eickwort, J., Smith, J., Hubbard, W., and Coyle, D. (2020). Forest health diagnostics Facebook page: Impact and natural resource programming implications. *Journal of Extension, 58*(3). https://tigerprints.clemson.edu/joe/vol58/iss3/25/

Fain, G. B., Gilliam, C. H., Tilt, K. M., Olive, J. W., & Wallace, B. (2000). Survey of best management practices in container production nurseries. *Journal of Environmental Horticulture, 18*(3), 142–144.

Hall, C. R., Hodges, A. W., Khachatryan, H., Palma, M. A. (2020). Economic contributions of the green industry in the United States in 2018. *Journal of Environmental Horticulture, 38*(3), 73–79.

Knudson, William A., & Peterson, Christopher H. (2012). *The economic impact of Michigan's food and agriculture system* (Working Paper No. 01-0312). Michigan State University Product Center.

National Agricultural Statistics Service, United States Department of Agriculture. (2007a). *Agricultural chemical usage 2006 nursery and floriculture summary.* United States Department of Agriculture Economics, Statistics and Market Information System. http://usda.mannlib.cornell.edu/MannUsda/viewDocumentInfo.do?documentID=1570

National Agricultural Statistics Service, United States Department of Agriculture. (2007b). *Nursery crop summary 2006.* United States Department of Agriculture Economics, Statistics and Market Information System. http://usda.mannlib.cornell.edu/MannUsda/viewDocumentInfo.do?documentID=1115

Building Extension Capacity through Internal Grants: Evaluation of a Mini-Grant Program

LENDEL K. NARINE[1], CRISTIAN MEIER[1], AND BRIAN HIGGINBOTHAM[1]

AUTHORS: [1]Utah State University.

Abstract. Acquiring external grants can seem out of reach for Extension professionals, especially early-career professionals. While Cooperative Extension provides opportunities to assist professionals in the grant writing process, Utah State University (USU) Extension facilitates an internal mini grant program to build professionals' capacity to apply for external funds. Using survey data from USU Extension professionals, our study sought to evaluate the processes and outcomes of the internal mini grant program. Our results provided recommendations to improve the program. Our study provides insights that can assist other institutions seeking to implement their own internal mini grant program.

INTRODUCTION

Cooperative Extension disseminates evidence-based information to the public to fulfill the land grant mission. While institutions facilitate innovative approaches to funding county-level programs (e.g., fee-based programs; Pellien, 2016), Extension county professionals (i.e., faculty and agents) must seek grants to fill funding gaps. However, acquiring grants, which is described as an important area of performance in a roadmap for excellence in Extension (Saunders & Reese, 2011), may seem out of reach for Extension professionals. There are specific factors that have been shown to influence grant awards, including (a) the number of proposals submitted, (b) the number of grant awards available, (c) participation in grant writing training, and (d) the size of the project team (Cole, 2006; Sisk, 2011).

For new Extension professionals, it takes time and resources (e.g., social capital and grant writing training) to develop the collaborative teams often necessary to acquire large external funding awards. Grant funding also impacts scholarly output. For example, one study found higher levels of grant funding were associated with increased publications (Kim et al., 2019), which further highlights the importance of grant funding. Extension can support county professionals by providing professional development opportunities to strengthen their abilities to pursue and receive grant awards.

Many Extension organizations already provide a number of opportunities to build professionals' capacity to apply for grant funding, such as grant writing trainings and mentoring programs. Yet, there is reason to believe that the act of applying for funding itself is a catalyst for collaboration among Extension professionals (Gould & Ham, 2002). The use of internal grants is one strategy for building the capacity of professionals to promote collaboration while also improving Extension professionals' experience in writing grant applications. In 2014, Utah State University (USU) Extension implemented an internal grant program, referred to as Extension mini grants.

Yearly mini grants have varied in the number of awards made (17 to 59 awards) and the total amount awarded each cycle (approximately $162,000 to $543,000), with a total investment to date of more than $2.7 million. Extension professionals can apply for these internal grants once a year to fund new and innovative Extension programs. Using two grant ceiling amounts based on the scope of the project (one county vs. multiple counties), the mini grant process provides clear instruction guidelines and is blind-reviewed by a peer panel of Extension professionals. The application and review process for Extension mini grants mimics the general process of applying for external grants. The primary goal of the mini grant program is to fund programs that improve the lives of Utah residents. Secondary goals are to build Extension professionals' capacity to apply for external funds, increase collaboration, and provide seed funding for innovative programs that may lead to external funding awards.

USU Extension issues a call for proposals once a year, and the application and selection process has remained relatively the same since the start of the program in 2014.

While the total number of awards and the maximum value of individual awards varies annually based on available administrative funds, the mini grants are an important resource available to Extension professionals. However, the internal mini grant program has never been evaluated to understand the return on investment that occurs for both the institution and Extension professionals, nor have there been any published research studies that describe the benefits of such a program. Therefore, this research-in-brief assesses the processes and outcomes of USU Extension's internal mini grant program.

This study adopts a summative evaluation design to determine the mini grant program's return on investment. Returns on investment can take the form of societal improvements and/or benefits to stakeholders based on measurable program outcomes. Therefore, summative evaluation determines the extent to which resources (i.e., investments) were used effectively and efficiently to achieve the program's intended benefits (Rossi et al., 2004). Results of a summative evaluation can assist planners in decisions about program continuation. In this context, proxy indicators are used to determine the return on investment of the mini grant program. These indicators broadly relate to Extension professionals' grant-writing competencies, secured external funding, and academic outcomes attributed to the mini grant program. Results can provide other Extension organizations with information about the mini grant program and potential outcomes of such a program.

PURPOSE AND OBJECTIVES

The purpose of this study was to evaluate the application processes and outcomes of the mini grant program at USU Extension. Objectives were to: (a) Describe Extension professionals' perceptions of the eligibility requirements for mini grants; (b) rank factors influencing the mini grant review and selection process; (c) determine the number of journal papers, conference submissions, factsheets, videos, e-courses, and external funding awards acquired as a direct result of the mini grant program; (d) describe the competencies gained by Extension professionals due to writing a mini grant proposal; and (e) understand what improvements could be made to the program to meet the needs of Extension professionals. Objectives (a), (b), and (e) relate to a formative evaluation of the mini grant program, while objectives (c) and (d) relate to the summative evaluation.

METHODS

This study followed a cross-sectional descriptive design and primary data were gathered from USU Extension professionals. The target population was all Extension professionals who were awarded at least one mini-grant between 2014 and 2019. A sampling frame was created from internal data provided by Extension administration. The sampling frame consisted of 103 Extension professionals ($N = 103$). With a census attempted, the response rate was 80% ($n = 82$). Each of the 82 professionals responding to the survey attained between one to two mini grants on average between 2014 and 2019 (M = 1.58, SD = 1.00).

Data were gathered in June of 2020 using an online questionnaire administered through Qualtrics. A panel of experts at USU Extension reviewed the questionnaire for face validity. A survey invitation was sent to the target population using Qualtrics. We tracked responses in Qualtrics and sent reminders to professionals who did not complete the survey in one-week intervals. The Associate Vice President for USU Extension sent two reminder emails to professionals of the target population. Data collection lasted three weeks following the initial survey invitation. The researcher-developed questionnaire was designed to gather data on pre-defined outcome indicators of the mini grant program. Leadership at USU Extension communicates the desired outcomes of a mini grant in annual requests for proposals. These include conference papers, journal articles, impact reports, short courses, and, eventually, external funding. The final questionnaire consisted of four sections: (a) professional appointment, (b) grant activity, (c) process evaluation, and (d) outcome evaluation. Extension professionals were also asked to comment on their experiences with the mini grant program via an open-ended question.

The process evaluation focused on two main areas: (a) Extension professionals' perceptions of the eligibility requirements for a mini grant and (b) factors influencing the mini grant review and selection process. The outcome evaluation focused on administratively defined outcome indicators of the mini grant program. These were: (a) Internal collaborations, (b) external collaborations, (c) journal articles, (d) conference papers, (e) factsheets, (f) videos, (g) e-courses, and (h) external funding. Extension professionals were asked to indicate the extent to which their mini grant(s) contributed to changes in each outcome.

Objective (a) was addressed using descriptive frequencies to rank perceived eligibility requirements. For factors influencing the review and selection process (objective b), respondents were asked to rank six pre-defined factors using a rank-order question format in Qualtrics. Respondents ordered the six items based on their rank preference, which resulted in a score between 1 (first rank) and 6 (last rank) for each item. Then, a repeated measures ANOVA was used to determine if there was a statistically significant difference between priority rankings. The null hypothesis was rejected at $p < 0.05$. For post-hoc analyses, we conducted a series of pairwise comparisons with a Bonferroni adjustment to p-values. Extension professionals at USU Extension typically progress through a tenure-track system. Therefore, results corresponding to objectives (a) and (b) were assessed by tenure status to examine the differences in perceptions between early-career faculty and others with respect to grant requirements and priorities.

Objective (c) was addressed using descriptive analysis (i.e., sum and means of outcomes within groups). First, a Q-Q plot was used to identify outliers in self-reported outcomes. Extreme values were removed from the dataset using the interquartile range method (IQR). However, due to the large variance in self-reported external funding attributed to mini grants across the sample, the mean value for external funding was supported with quartiles to further illustrate the data spread and median (i.e., 50th quartile). The total number of outcomes (e.g., journal articles, conference papers, etc.) were divided by the total number of grants across the sample to derive mean outcomes per mini grant. Outcomes of the mini grant program were reported by tenure status and program area. Frequencies were used for objective (d) to describe competency gained by Extension professionals through the mini grant program.

Finally, to analyze qualitative data for the open-ended question related to objective (e), we utilized a two-step coding procedure (Saldaña, 2016) where data was coded as categories emerged (i.e., pattern coding). The data was first coded by one member of the research team, then was reviewed by a second member of the team. If there was a disagreement in coding, the two coders discussed the code and reached an agreement on the suitability of the code.

As a retrospective study, there are two major limitations to our project. It should be noted that all data provided by respondents are approximations and are based on their ability to self-report the ripple effects of funding from the mini grant program. As a result, there may be recall bias in self-reported estimations, particularly with respect to external funding attained due to mini grants. Another limitation is the use of a cross-sectional (non-experimental) design. We are unable to determine a true causal relationship between the acquisition of a mini grant and eventual realization of the described outcomes with respect to academic productivity.

RESULTS

SAMPLE CHARACTERISTICS AND GRANT ACTIVITY

More than half the number of respondents were tenured Extension professionals (51%); 26% were untenured, and 23% were categorized as "other" (e.g. 4-H coordinators, administrators). Most Extension professionals listed agriculture and natural resources as their primary program area (49%); 18% listed family and consumer sciences, 10% listed 4-H and youth development, and 2.4% listed economic development. However, 21% were unable to list their primary program area due to assignment splits.

Table 1 shows the level of grant activity by program area from 2014 to 2019. Overall, the majority of mini grant funding ($994,801) was acquired by professionals in agriculture and natural resources, while the least ($157,045) was acquired by professionals in 4-H and youth development. However, family and consumer sciences professionals acquired the most grants on a per capita basis (1.87) compared to professionals in other departments (1.63). Extension professionals in agriculture and natural resources acquired individual grants of higher value ($26,178) compared to professionals in family and consumer sciences ($24,245) and 4-H and youth development ($19,630).

PROCESS EVALUATION

Table 2 shows respondents' perceptions of various aspects of the application process. Results are presented by tenure status to assess the perceptions of early-career Extension professionals in comparison to others. Overall, most professionals (79%) thought a first-time grant applicant should secure a mentor when writing their proposal; 66% thought all proposals should include a collaboration between county Extension professionals and campus professionals; and 61% thought proposals from junior professionals should be prioritized over others. While these results were somewhat consistent across tenured and untenured professionals, there were differing opinions on one requirement: more than half the number of untenured professionals (60%) indicated grants

Table 1. Grant Activity by Program Area

Program Area	n	Total		Mean	
		Number of Grants	*Value of Grants*	*Number of Grants*	*Value of Grants*
4-H & Youth Development	8	13	$157,045	1.63	$19,630
Family & Consumer Sciences	15	28	$363,679	1.87	$24,245
Agriculture & Natural Resources	38	62	$994,801	1.63	$26,178
Other	17	20	$292,651	1.17	$17,214
Sample total (2014–2019)		123	$1,808,176		
*Actual total (2014–2019)		182	$2,383,571		

Note. Actual values provided by USU Extension administration.

Table 2. Perceptions of Eligibility Requirements

Rank	Requirement	Tenured (n = 42)		Untenured (n = 21)		Other (n = 18)		Overall (n = 82)	
		Yes	No	Yes	No	Yes	No	Yes	No
1	A first-time grant applicant should secure a mentor when writing her/his grant proposal	77	23	80	20	83	17	79	21
2	Proposals should include a collaboration between county Extension professionals and campus faculty	67	33	65	35	67	33	66	34
3	Proposals from junior professionals should be prioritized over others	56	44	60	40	72	28	61	39
4	Grants should only be awarded to proposals with a clear potential for external funding	36	64	60	40	72	28	42	58

Table 3. Factors for Consideration in Grant Review Process

Overall Rank	Proposals…	Mean Rank (SD)			
		Tenured	Untenured	Other	Overall
1	…with collaboration between campus faculty and county professionals	2.79 (1.66)	2.35 (1.57)	2.11 (1.57)	2.52 (1.62)
1	…that can lead to significant impacts	2.77 (1.69)	2.70 (1.34)	2.94 (1.63)	2.79 (1.58)
2	…with high scores from reviewers	3.54 (1.67)	3.80 (1.94)	3.89 (1.71)	3.69 (1.73)
2	…with a clear plan to secure external funding	3.87 (1.42)	3.30 (1.34)	3.78 (1.22)	3.70 (1.36)
2	…with high relevance to Extension programs	3.59 (1.65)	4.10 (1.41)	4.06 (1.73)	3.83 (1.61)
3	…from junior campus faculty and professionals	4.44 (1.65)	4.75 (1.55)	4.22 (1.59)	4.47 (1.60)

Note. Overall rank denotes statistically significant differences between priority rankings for the overall sample based on a repeated measures ANOVA with Bonferroni-adjusted pairwise post hoc tests.

should only be awarded to proposals with a clear potential for external funding. This finding points towards one of the main goals of the mini grant program; it appears untenured professionals believe the program should be used as seed funding to attain external grants. In contrast, only 36% of tenured professionals thought this should be a requirement for mini grant funding.

Table 3 shows respondents' perceptions on priority factors that should influence the grant review and selection process. Results of a repeated measures ANOVA indicated there was a statistically significant difference in priority rankings for the overall sample (Greenhouse-Geisser $F_{(4.22, 320.87)} = 13.17$, $p < 0.01$). However, results showed there were no statistical differences in the interaction between priority rankings

and groups (tenure vs. untenured vs. other). This suggests the overall ranking holds for all professionals regardless of tenure status. Overall, professionals thought the top factors that should be weighted the most in the review process were (1) proposals that included a collaboration between campus professionals and county professionals and (2) proposals that can lead to significant impacts.

OUTCOME EVALUATION

Most respondents (96%) strongly agreed or agreed their mini grants led to an increase in their collaborations with other professionals within USU Extension. Slightly less (74%) strongly agreed or agreed that their mini grants led to an increase in their collaborations with professionals/staff outside USU. In addition, 65% either strongly agreed or agreed their mini grants led to an increase in their peer-reviewed journal publications, and 87% strongly agreed or agreed it led to an increase in their conference paper submissions.

Table 4 provides a summary of the outcomes of the mini-grant program from 2014 to 2019. Overall, mini grants contributed mostly to factsheets (259), conference papers (239), and videos (217). It also led to a total of approximately $16 million in external funding. Other noteworthy outcomes of the mini grant program were contributions to journal publications (90) and e-courses (17) from both tenured and untenured professionals.

Results indicated that, on average, one mini grant led to one journal paper (M = 0.82, SD = 1.01), two conference papers (M = 2.21, SD = 2.12), three factsheets (M = 2.68, SD = 5.38), one video (M = 1.39, SD = 4.35), and $138,469 in external funding (M = $138,496.32, SD = $290,537.27: 25th Quartile = $0.00, 50th Quartile = $20,000, 75th Quartile = $78,034).

Professionals were asked to self-assess the competencies they developed as a result of writing a mini grant proposal. Table 5 shows competencies ranked based on the frequency of "Yes" responses to each item. More than half the number of professionals indicated they developed the competencies to seek collaboration with peers, understand the grant writing process, and create a grant budget because of the mini grant program.

Finally, Extension professionals were asked if they had any recommendations to improve the mini-grant program in an open-ended question. Extension Professionals provided a variety of recommendations to improve the mini-grant program (n = 46). Response themes were: (a) provide a mentor, training, or other resources to improve grant writing skills; (b) simplify the grant application process; (c) prioritize county-level Extension work instead of campus research; (d) give preference to junior faculty seeking grants; (e) reduce emphasis on attaining external funding from the mini grant; (f) provide training for reviewers; and (g) encourage applications that aim to pilot innovative programs.

Table 4. Outcomes by Program Area and Tenure Status

Factor	Level	n	Total					
			Journal	Conference	Factsheets	Videos	E-courses	External funding ($)
Tenure Status	Tenured	39	59	141	99	151	12	9,358,367
	Untenured	20	18	52	92	20	3	1,614,000
	Other	18	13	46	68	46	2	5,384,105
Program Area	4-H	8	5	21	34	27	3	1,828,822
	FCS	15	19	44	44	117	8	4,398,314
	AG/NR	38	51	126	120	26	2	4,786,336
	Other	16	15	48	61	47	4	5,343,000
Overall Total		77	90	239	259	217	17	16,356,472

Table 5. Competency Gained From the Mini Grant Program

Rank	Did the Extension mini-grant program help you to better understand...	%			
		Yes	Unsure	No	Knew before
1	how to seek collaboration with peers?	57	5	5	33
2	the general grant writing process?	51	4	4	42
2	how to create a grant budget?	51	1	4	44
3	how to create a project evaluation plan?	48	13	9	30
4	how to write a concise problem statement?	47	7	5	42
5	how to describe the project methodology?	46	4	8	43
5	how to disseminate grant results?	46	13	5	36
6	how to write project proposal goals?	44	7	5	44
7	how to manage a grant budget?	43	4	8	46

CONCLUSIONS AND RECOMMENDATIONS

Our study sought to evaluate the internal mini grant program at USU Extension. Results pointed towards several recommendations that can improve the program. This study can help guide other institutions seeking to implement a similar program. First, faculty perceptions toward mini grant eligibility (i.e., objective a) pointed towards needs for mentoring to assist first-time applicants and collaboration on proposals between county and campus professionals. A majority of tenured and untenured professionals reported that mentoring should be secured when writing a grant application. This was also mentioned in respondents' comments. We also recommend providing additional trainings in the form of webinars or workshops to assist new and early career professionals in grant preparation, which supports the findings of previous research (Cole, 2006; Sisk, 2011). Specifically, a professional development webinar can be offered to applicants suggesting tips for writing a better mini grant proposal and addressing common mistakes to avoid.

The results show that administrators can consider weighting specific factors differently in the review and selection of mini grant proposals (i.e., objective b). For example, results show collaborations between county and campus faculty should be prioritized in mini grant proposals. Future requests for mini grant proposals could emphasize the importance of collaborations; this may include promoting collaborations between early career professionals and others as a mentorship and capacity-building activity. During the review process, proposals that demonstrate collaborations between campus and county professionals could receive special consideration when making funding decisions. Another important factor in the review process was potential impact; respondents thought the mini grant proposal program should articulate its potential to generate significant impacts. This suggests the need for a robust evaluation plan as a core component of mini grant proposals. An effective evaluation plan could describe a need, problem statement, and intended outcomes and long-term impacts of the project.

An examination of outcomes of the mini grants program showed several noteworthy findings (i.e., objective c). Outcomes associated with mini grants are important for early career Extension professionals as they work towards tenure and promotion (e.g., peer reviewed articles). However, results also showed some outcomes were just as important for tenured Extension professionals (e.g., external funding). This suggests the mini grant program has differing outcomes for pre-tenure and tenured Extension professionals, and, as such, the resulting tangible value of participating in the mini grant program may be beneficial to all Extension professionals, regardless of tenure status.

Finally, results from the qualitative analysis further confirm the importance of collaboration in mini grant proposals (i.e., objective e). Findings indicate more than half of respondents thought that collaboration between county and campus Extension professionals should be prioritized in the grant review process. Similarly, most respondents also said the mini grant process contributed to their skills in seeking collaboration with colleagues (i.e., objective d). Clearly, the facilitation of collaboration between professionals should be an important aspect of the mini grant program and can be encouraged in the call for proposals.

These findings will inform changes to the mini-grant program at USU Extension. Other universities can use the results of this study to inform the development and implementation of their own internal grant program. Implementing an internal grant program in Extension may have a high potential for return on investment while building the capacity of professionals to be successful in seeking external grant funding and boosting their academic productivity and competencies.

REFERENCES

Cole, S. S. (2006). Researcher behavior that leads to success in obtaining grant funding: A model for success. *Research Management Review*, 15(2), 1–16.

Gould, R., & Ham, G. (2002). The integration of research and extension: A preliminary study. *Director*, 62(67), 42.

Kim, S., Li, B., Rosoklija, I., Johnson, E. K., Yerkes, E., & Chu, D. I. (2019). Federal research funding and academic productivity in pediatric urology: From early career to research independence. *Journal of Pediatric Urology*, 15(3), 233–239.

Pellien, T. (2016). Building 4-H program capacity and sustainability through collaborative fee-based programs. *Journal of Extension*, 54(2). https://tigerprints.clemson.edu/joe/vol54/iss2/19

Rossi, P., Lipsey, M., & Freeman, H. (2004). *Evaluation: A systematic approach* (7th ed.). Sage.

Saldaña, J. (2016). *The coding manual for qualitative researchers*. Sage.

Saunders, K. S., & Reese, D. (2011). Developing a roadmap for excellence in Extension. *Journal of Extension*, 49(3). https://tigerprints.clemson.edu/joe/vol49/iss3/24

Sisk, A. (2011). How critical is training? Impact of a semester-long proposal writing course on obtaining grant funding. *Journal of the Grant Professionals Association*, 9(1), 66–77.

JOURNAL OF
Extension

Research in Brief

Volume 60, Issue 2, 2022

Results of a Rural Traffic Calming Event to Promote Physical Activity

MARISSA J. SPEAR[1], BRETT ROWLAND[1], JESSICA VINCENT[2], TYLER BROWN[2], ADDIE WILSON[2], CAITLIN PALENSKE[2], PEARL A. MCELFISH[1], CHRISTOPHER R. LONG[1], JESSICA PRESLEY[3], AND LAURA E. BALIS[4]

AUTHORS: [1]University of Arkansas for Medical Sciences Northwest. [2]University of Arkansas System Division of Agriculture. [3]University of Arkansas for Medical Sciences. [4]Louisville Center, Pacific Institute for Research and Evaluation.

Abstract. This article describes how community need was addressed through a traffic calming pop-up event in rural Arkansas. The event was conducted on routes connecting a neighborhood, two schools, and a municipal park. A brief survey assessed safety concerns of parents and/or guardians related to children walking or biking to school. Prior to the event, parents/guardians reported it was not safe for their children to walk or bike to school; however, the majority agreed the event made the area safer. Small-scale traffic calming events can provide evidence to stakeholders that built environment changes are an important childhood obesity prevention strategy in rural Extension work.

INTRODUCTION

Evidence demonstrates that child participation in physical activity reduces rates of obesity and other chronic conditions and improves academic achievement (U.S. Department of Health and Human Services, 2018). Improving access to active transportation, such as walking or biking, increases rates of physical activity in children (Ward et al., 2015). Safe Routes to School programs have been successful in encouraging children to walk and bike to school (Stewart et al., 2014). However, in recent years, there has been a decrease in children walking or biking to school, with parents citing barriers such as distance, traffic-related danger, and crime (Jones & Sliwa, 2016). Childhood obesity has been linked to numerous health problems, including type 2 diabetes and hypertension, and is a strong predictor of adult obesity (Pandita et al., 2016; Simmonds et al., 2016). Therefore, primary and secondary prevention of childhood obesity is a critical step in addressing the adult obesity epidemic in the United States.

The Centers for Disease Control and Prevention (CDC) implemented the High Obesity Program (HOP) to address health issues in counties with an adult obesity rate of over 40% through increasing access to healthy foods and providing safe and accessible places for physical activity. The CDC funded 15 Land-Grant University Cooperative Extension Systems to complete this work, aligning with Extension's

National Framework for Health and Wellness and recent addition of physical activity as a focus area (Braun et al., 2014; CDC Division of Nutrition, Physical Activity, and Obesity, 2020). Program activities include collaborating with community partners to connect everyday destinations that could be linked via active transportation, including homes, schools, parks, and other community locations. Each awardee is required to evaluate the effectiveness of their strategies within these program activities.

PURPOSE AND OBJECTIVES

In 2018, the Arkansas Cooperative Extension Service received the five-year HOP cooperative agreement to implement obesity prevention strategies in the Arkansas Delta region. The Arkansas Delta Region Obesity Project (ArDROP) is available to five counties: Chicot, Lee, Mississippi, Philips, and St. Francis. Mortality rates in the Delta region from chronic diseases are 10% higher than non-Delta counties and 20% higher than the national rate (Felix & Stewart, 2005). Despite a steady decrease in mortality nationally, the rate of decline in the Delta is much slower (Cosby & Bowser, 2008).

St. Francis County has the highest adult obesity rate in Arkansas (43.2%), making it a critical location for obesity prevention strategies (Arkansas Department of Health, 2018). Forrest City is the St. Francis County seat and largest city, with a population of 15,371 (Missouri Census Data Center, 2020). More than 2 in 5 (41.9%) students in the Forrest City school district are overweight or obese (ACHI, 2019). The high rate of adult and childhood obesity, the effectiveness of safe routes to school interventions, a strong relationship with city officials, and the support of an active community coalition motivated the ArDROP team to begin work on safe routes to school in Forrest City. Prior work within the community included helping the city secure Transportation Alternatives Program (TAP) funding, conducting healthy food assessments in local food pantries, and conducting walkability assessments in other areas of the city. This article describes the steps taken by the ArDROP team to increase access to physical activity in the community.

INTERVENTION APPROACH

The Mayor of Forrest City first identified several areas of need, which initiated ArDROP's activities to encourage the development of safe routes to school. ArDROP chose the area near Oak Avenue and Ash Street, because it contains a residential neighborhood, two schools, and a municipal park. The ArDROP team received training on conducting walkability assessments from CDC subject matter experts. ArDROP then conducted a walkability assessment of the identified area using a tool adapted from CDC's Walkability Audit Tool (Dannenberg et al., 2005) and North Carolina's Shape Your World Walkability Checklist (Move More Walk Now, 2011). The results of the walkability assessment demonstrated that the area was not pedestrian-friendly. Specifically, the area lacked sidewalks, crosswalks, connectivity to the community, safety features, and an attractive appearance. Previous rural Extension work has also cited inadequate signage, lack of sidewalks and crosswalks, and low pedestrian safety as significant barriers to walking and biking for rural residents (Jensen et al., 2019).

Based on the results of the walkability assessment, ArDROP developed plans to conduct a "traffic calming pop-up" event in the area to demonstrate the feasibility of making small built environment changes that could encourage active transportation. Traffic calming pop-ups are a strategy of tactical urbanism in which simple, temporary changes to the built environment are tested, including traffic cones, signs, and paint, in order to demonstrate effective ways a city government can make permanent changes to improve the built environment (Safe Routes to School National Partnership, 2017; Trailnet, 2016). The ArDROP team conducting the pop-up event included five program associates, the lead evaluator, and the principal investigator.

The ArDROP team rerouted traffic to turn Oak and Ash into one-way streets. The team transformed Oak Avenue into a one-way street going west (toward the schools) and Ash Street into a one-way street going east (away from the

schools) (Figure 1). For a detailed list of the methods and purpose of the temporary built environment changes used in Figure 1, see Table 1. The event occurred from 2 PM to 5 PM on a school day in September 2019.

EVALUATION METHODS

The team evaluated the event through observation during the event and a brief post-event survey. The ArDROP team used paper, pencil, and a mobile app to count the number of pedestrians, cyclists, and cars that drove down both Oak and Ash throughout the duration of the event. This count was an approximate measure conducted for process evaluation (i.e., to provide data to stakeholders on how many community members were exposed to the pop-up event), and, thus, the ArDROP team did not determine inter-rater reliability.

School staff were asked to distribute the survey to students at the local middle and elementary schools after the traffic calming pop-up event (Figure 1 does not depict the elementary school). School staff asked the students to take the survey home, have a parent or guardian complete the survey, and return the completed survey the following day. Passive consent was obtained through return of the completed survey. The eight-item survey asked students' parents/guardians to report how often their child walked or biked to school, whether it was safe for their child to walk/bike to school, concerns about letting their child walk/bike to school, whether or not they thought the event improved safety, and whether or not permanent changes would increase the likelihood of allowing their children to walk or bike to school. The ArDROP team collected completed surveys from the schools two weeks after they conducted the pop-up event. The team entered survey data into Microsoft Excel for analysis. This evaluation was determined to be exempt by the University of Arkansas for Medical Sciences Institutional Review Board.

RESULTS

The ArDROP team counted approximately 140 vehicles, 24 pedestrians, and one cyclist throughout the duration of the event. All vehicles and pedestrians adhered to the temporary built environment changes throughout the event.

A total of 49 parents/guardians responded to the survey. Summary statistics are provided in Table 2. According to survey responses, the majority of parents/guardians reported that their children never walk or bike to school (85.4% and 100%, respectively). Three-quarters (76.6%) of parents/guardians disagreed with the statement that it is safe for their child to walk/bike to school prior to the event. Parents/guardians listed several concerns about their children walking or biking to school, including distance, lack of sidewalks/crosswalks, speed and amount of traffic, and crime. The majority (85.3%) agreed that the changes to the area made

Results of a Rural Traffic Calming Event to Promote Physical Activity

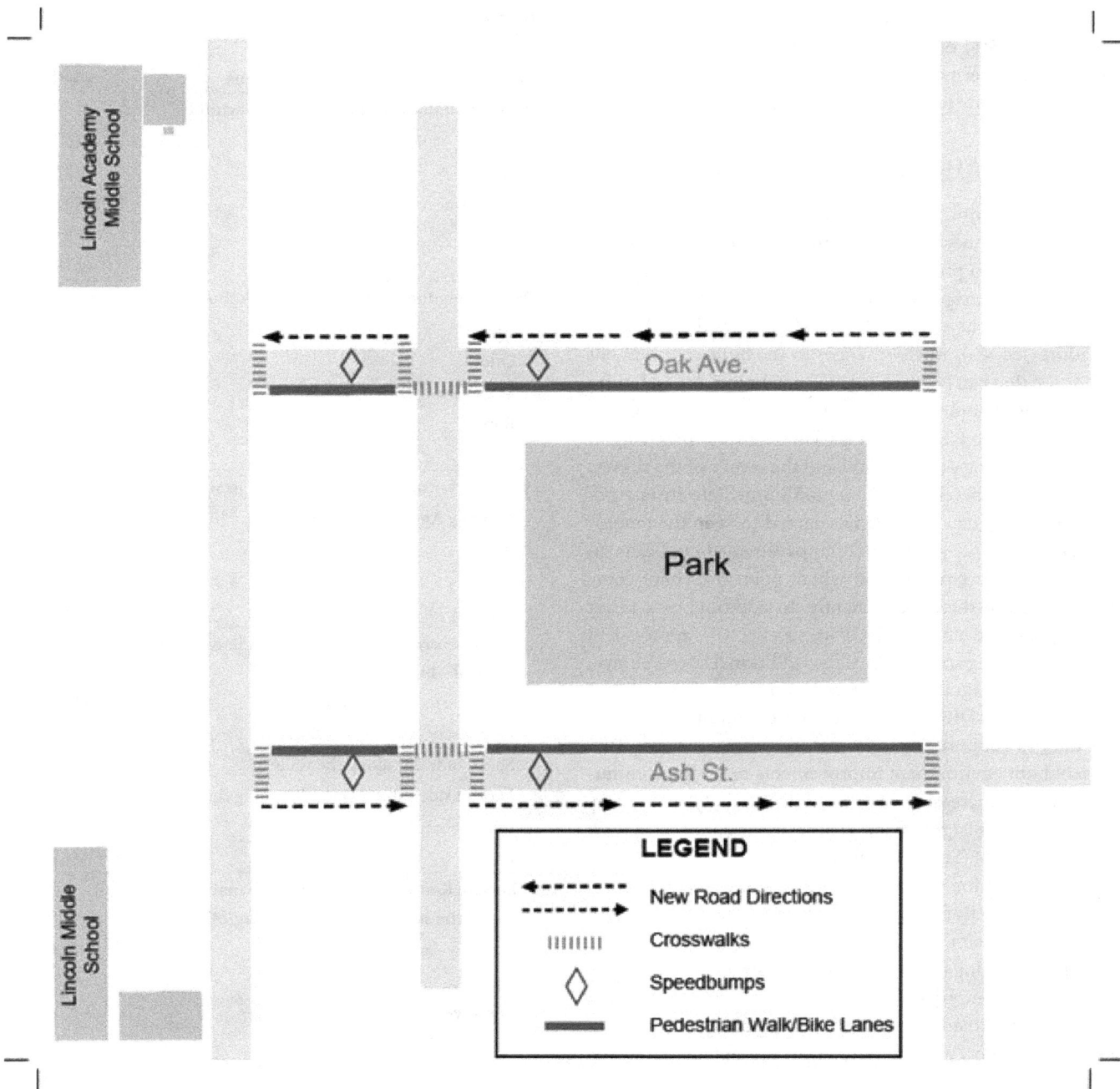

Figure 1. Map of temporary built environment changes.

Table 1. Methods and Purpose of Temporary Built Environment Changes

Methods of Environment Changes	Purpose
Added dedicated walking/biking lanes to Oak and Ash using temporary pavement tape and yellow footprint stickers	To account for the lack of sidewalks
Added crosswalks at all four intersections using temporary pavement tape	To account for the lack of crosswalks
Added four temporary speedbumps on Oak and Ash	To slow traffic travelling to and from the schools
Added temporary signage to stop signs in affected area	To indicate the newly designed one-way streets
Placed traffic cones near new crosswalks and pedestrian lanes	To aid in pedestrian visibility

it safer for their children to walk/bike to school. More than one-third (36.8%) of parents/guardians agreed that making these changes permanent would increase the likelihood of allowing their kids to walk or bike to school.

IMPLICATIONS FOR PUBLIC HEALTH

The traffic calming pop-up event provided preliminary support for the necessity of permanent built environment improvements to promote physical activity. Two of the primary concerns respondents identified—lack of sidewalks/crosswalks and the speed of traffic—are issues that are feasibly addressed with micro-changes to the built environment (i.e., those that can be implemented at a lower cost and more rapidly than complex infrastructure improvements) (Community Preventive Services Task Force, 2016). Based upon the traffic calming pop-up event and the results of the survey, ArDROP and the city implemented immediate infrastructure improvements. City officials agreed to keep the temporary crosswalks in place until the pavement tape began to detach from the pavement, at which point they have committed to paint them permanently. In addition, as a result of the pop-up event, ArDROP received CDC approval to purchase flashing speed limit signs, additional speed bumps, and pedestrian signage to slow traffic on Oak and Ash. Taken together, the ArDROP team received community feedback and data to support the need for permanent changes. Permanent built environment improvements make the community safer and increase access to physical activity, which may encourage parents/guardians to allow their children to walk or bike to school (Stewart et al., 2014).

One limitation of this study was that the ArDROP team did not observe traffic in the area prior to the pop-up event; therefore, the team was unable to compare use prior to the event to use during the event. Another limitation was the low number of parent/guardian responses to the survey. Surveys were delivered to administration offices at both schools, but it is unknown how many children were present on the day the surveys were distributed or whether every student was given a survey to take home to their parents/guardians; therefore, a precise response rate could not be calculated.

The majority of parents/guardians reported distance as a concern; however, the ArDROP team did not have data detailing the distance respondents lived from the school. Therefore, further work should investigate what distance is feasible for parents/guardians to allow children to walk or bike to school in order to make safe routes projects and traffic calming measures effective. Parents/guardians also reported crime as a concern, which was not specifically addressed with temporary changes and may account for the proportion of respondents who would *not* be more likely to allow their children to walk or bike to school. Before crime prevention can be addressed, stakeholders need to conduct

Table 2. Parent/Guardian Attitudes and Concerns about Active School Transportation, n=49

Survey Items And Response Options	no. (%)
How often does your child walk to school?	
Always	3 (6.3)
Often	0 (0)
Sometimes	0 (0)
Rarely	4 (8.3)
Never	41 (85.4)
How often does your child bike to school?	
Always	0 (0)
Often	0 (0)
Sometimes	0 (0)
Rarely	0 (0)
Never	47 (100)
It is safe for my child to walk/bike to school.	
Strongly Agree	1 (2.1)
Agree	10 (21.3)
Disagree	18 (38.3)
Strongly Disagree	18 (38.3)
What concerns do you have about letting your child walk/bike to school?[a]	
Distance to/from school	34 (69.4)
Amount of traffic along the route	24 (49.0)
Speed of traffic along the route	22 (44.9)
Lack of sidewalks/crosswalks along the route	18 (36.7)
Crime along the route	15 (30.6)
Other	6 (12.2)
The Safe Routes to School pop-up made the area around the school safer for walking/biking.	
Strongly Agree	3 (8.8)
Agree	26 (76.5)
Disagree	1 (2.9)
Strongly Disagree	4 (11.8)
If the temporary changes were made permanent, it would increase the likelihood that I would walk/bike to school with my child or let my child walk/bike to school.	
Strongly Agree	1 (2.6)
Agree	13 (34.2)
Disagree	14 (36.8)
Strongly Disagree	10 (26.3)

Note. Number of responses varies by item due to participant non-response. Percentages may not total 100 due to rounding.
[a]Multiple responses were allowed for this item.

Journal of Extension

Volume 60, Issue 2 (2022)

additional assessments to determine the types of crime and environmental elements that impact personal safety—such as visibility, obstruction, and areas of concealment (Cozens et al., 2005). This knowledge can guide future efforts to prevent crime through environmental design (e.g., natural surveillance and access control or maintenance) (Cozens et al., 2005).

CONCLUSION

Physical activity promotion and environmental change are both recent focuses of Extension work (United States Department of Agriculture, 2014). This case study may serve as an example for other Extension practitioners. Safe Routes to School National Partnerships recommend pop-ups (Safe Routes to School National Partnership, 2017); however, as of yet, pop-ups have been unreported in the Extension literature. These brief feasibility studies can be used within the health promotion process to work with community coalitions, identify needs, and select, test, implement, and evaluate evidence-based interventions. This case study demonstrates how a brief traffic calming pop-up event was used to make permanent built environment changes in a short period of time. The ArDROP team will continue working with community-driven coalitions like the one in Forrest City to conduct traffic calming pop-up events in other counties and to demonstrate to city governments the feasibility of improving the built environment. This article demonstrates simple strategies for encouraging communities to increase physical activity by walking and biking to school. Changes to the built environment are easily replicable in Extension work in rural areas and can be a critical first step in reducing childhood obesity.

DISCLAIMERS

We have no conflicts of interests to declare. No copyrighted materials, surveys, instruments, or tools were used in this work. This work was funded by a High Obesity Program (HOP) award (no. NU58DP006561) from the Centers for Disease Control and Prevention. The findings and conclusions in this report are ours and do not necessarily represent the official positions of the funder.

ACKNOWLEDGMENTS

We would like to thank St. Francis County Extension Agent Cody Griffin; Phillips County Extension Agent Julie Goings; Jennifer Conner; Jordyn Williams; Diane Ayers and Sahra Kahin of the CDC; the Forrest City Police Department and City Council; and the Mayor of Forrest City, Cedric Williams.

REFERENCES

ACHI (2019). *Assessment of Childhood and Adolescent Obesity in Arkansas: Year 16 (Fall 2018-Spring 2019).* Arkansas Center for Health Improvement. https://achi.net/wp-content/uploads/2019/12/191205-Year-16-2018-19-State-BMI-Report.pdf

Arkansas Department of Health (2018). *Arkansas BRFSS 2018 County Estimates: Overweight and Obese Adults.* https://ssl-adh.ark.org/images/uploads/pdf/Obesity-Overweight_2018.pdf Braun, B., Bruns, K., Cronk, L., Fox, L. K., Koukel, S., LeMenestrel, S., Lord, L. M., Reeves, C., Rennekamp, R., Rice, C., Rodgers, M., Samuel, J., Vail, A., & Warren, T. (2014). *Cooperative Extension's national framework for health and wellness.* ECOP Health Task Force. https://www.aplu.org/members/commissions/food-environment-and-renewable-resources/CFERR_Library/national-framework-for-health-and-wellness/file

CDC Division of Nutrition, Physical Activity, and Obesity (2020). *High obesity program.* Centers for Disease Control and Prevention. https://www.cdc.gov/nccdphp/dnpao/state-local-programs/hop-1809/high-obesity-program-1809.html

Community Preventive Services Task Force (2016). *Physical activity: Built environment approaches combining transportation system interventions with land use and environmental design, Task Force Finding and Rationale Statement.* The Community Guide. https://www.thecommunityguide.org/sites/default/files/assets/PA-Built-Environments.pdf

Cosby, A. G., & Bowser, D. M. (2008). The health of the Delta Region: A story of increasing disparities. *Journal of Health and Human Services Administration, 31*(1), 58-71. https://pubmed.ncbi.nlm.nih.gov/18575148/

Cozens, P., Saville, G., & Hillier, D. (2005). Crime prevention through environmental design (CPTED): A review and modern bibliography. *Property Management, 23*(5), 328-356. www.doi.org/10.1108/02637470510631483

Dannenberg, A. L., Cramer, T. W., & Gibson, C. J. (2005). Assessing the walkability of the workplace: A new audit tool. *American Journal of Health Promotion, 20*(1), 39-44. www.doi.org/10.4278/0890-1171-20.1.39

Felix, H., & Stewart, M. K. (2005). Health status in the Mississippi River Delta region. *Southern Medical Journal, 98*(2), 149-155.

Jensen, K., Tifft, K., Winfield, T., Gunter, K., Karp, G., & John, D. (2019). Engaging residents in participatory photomapping and readiness conversations to address the rural obesogenic context. *Journal of Extension, 57*(5). https://archives.joe.org/joe/2019october/a1.php

Jones, S. E., & Sliwa, S. (2016). Peer reviewed: School factors associated with the percentage of students who walk or bike to school, school health policies and practices study, 2014. *Preventing chronic disease, 13*. https://doi.org/10.5888/pcd13.150573

Missouri Census Data Center (2020). Census 2010 Profile Report.

Move More Walk Now (2011). *Walkability checklist.* Shape Your World. https://movemorewalknownc.com/wp-content/themes/WalkNow/downloads/Walkability_Checklist.pdf

Pandita, A., Sharma, D., Pandita, D., Pawar, S., Tariq, M., & Kaul, A. (2016). Childhood obesity: Prevention is better than cure. *Diabetes, Metabolic Syndrome and Obesity: Targets and Therapy, 9*, 83-89. www.doi.org/10.2147/DMSO.S90783

Safe Routes to School National Partnership (2017). *Pop-Ups for safe routes to school: Using tactical urbanism to promote Safe Routes to School programs.* https://www.saferoutespartnership.org/sites/default/files/resource_files/pop-ups_for_safe_routes_to_school_0.pdf

Simmonds, M., Llewellyn, A., Owen, C. G., & Woolacott, N. (2016). Predicting adult obesity from childhood obesity: A systematic review and meta-analysis. *Obesity Reviews, 17*(2), 95-107. www.doi.org/10.1111/obr.12334

Stewart, O., Moudon, A. V., & Claybrooke, C. (2014). Multistate evaluation of safe routes to school programs. *American Journal of Health Promotion, 28*(supplement 3), S89-S96. www.doi.org/10.4278/ajhp.130430-QUAN-210

Trailnet (2016). *Slow your street: A how-to guide for pop-up traffic calming.* Plan4Health, trailnet, American Planning Association, and St. Louis Department of Health. http://www.onestl.org/media/site/documents/reports/bicycle-pedestrian-planning/SlowYourStreets_HowTo Guide_Final-v.2_reduced.pdf

U.S. Department of Health and Human Services (2018). *Physical Activity Guidelines for Americans, 2nd edition.* https://health.gov/sites/default/files/2019-09/Physical_Activity_Guidelines_2nd_edition.pdf

United States Department of Agriculture (2014). *Local school wellness policy implementation under the Healthy, Hunger-Free Kids Act of 2010.* Federal Register. https://www.govinfo.gov/content/pkg/FR-2014-02-26/pdf/2014-04100.pdf

Ward, S., Bélanger, M., Donovan, D., Caissie, I., Goguen, J., & Vanasse, A. (2015). Association between school policies and built environment, and youth's participation in various types of physical activities. *Journal of School Health, 85*(7), 423-432. www.doi.org/10.1111/josh.12273

JOURNAL OF
Extension

Ideas at Work

Volume 60, Issue 2, 2022

Working Together for Soil Health: Liberating Structures for Participatory Learning in Extension

Carol R. McFarland[1], Claire Friedrichsen[2], Haiying Tao[3], and Maren L. Friesen[1]

AUTHORS: [1]Washington State University. [2]University of Idaho. [3]University of Connecticut.

Abstract. Liberating Structures (LS) provide a user-friendly toolkit to shift group power dynamics and allow all stakeholders to contribute. We explored the novel use of LS in soil health extension to conduct high-engagement events with diverse stakeholders. Our goals were to promote social learning, networking, and to encourage innovation. Soil health themes emerged highlighting specific practices, and the necessity of addressing broader scope issues of education, economics, and policy. Participants reported increased knowledge of soil health, professional connections, and forecasted participation in soil-health-promoting activities. Participants also expressed a sense of community, expanded perspectives, and appreciation of the co-development process.

INTRODUCTION

A dominant strategy within Extension has been 'linear technology transfer,' a paradigm encouraging adoption of emergent agricultural practices though one-way communication and the standard presentation-by-expert-to-audience format (Klerkx et al., 2012; Rogers, 2003). Extension modalities are moving towards adoption of adult education strategies to enhance learning effectiveness (Bell & McAllister, 2021). These efforts include high-engagement activities and co-innovation processes focused on social learning between producers, crop advisors, conservationists, and researchers toward integrating all stakeholders' knowledge and experiences. Co-learning has been shown to create knowledge that is pragmatic, tangible, contextually relevant, actionable, and focused on fostering community (Bremer & Meisch, 2017; Jagannathan et al., 2020).

As outreach models transition, Extension professionals require strategies encouraging participant-driven innovation processes across diverse stakeholders that promote group equity, develop the innovative capacity of individuals, and foster innovation-promoting social networks (Klerkx et al., 2012). Wang et al. (2019) found that potential behavior change was increased when perceived risk for adoption was lowered and awareness of the benefits of conservation practices was enhanced, a phenomenon described in the social sciences as the expansion of the mental model. Our team identified Liberating Structures (LS) as proven and versatile tools to create environments with high levels of engagement for co-learning while supporting expansion of participants' mental models.

Liberating Structures (LS) is a user-friendly toolkit for developing events that build connections and foster innovation within complex systems. See *The Surprising Power of Liberating Structures: Simple Rules to Unleash a Culture of Innovation* by Lipmanowicz and McCandless (2014) for the LS menu. The theoretical and conceptual framework of LS is based on a history of educational engagement strategies including Socrates, Montessori, and more recently, complexity science and action inquiry (Singhal et al., 2020; Torbert, 1991). Using LS supports both equitable power distribution within a group and participant engagement in peer-to-peer inquiry, but retains enough power for leadership to enable progress toward group objectives (Torbert, 1991). These tools have been successfully employed in a variety of fields, including increasing engagement in university classrooms and improving healthcare standards (Singhal et al., 2020; Mallet & Rykert, 2018; Chumley & Magrane, 2011; Mahoney et al., 2016). The use of LS in agricultural Extension to conduct participatory events with diverse stakeholders can promote a shift toward social learning, networking, and co-innovation.

METHODS

As part of the Washington Soil Health Initiative, a LS format was used for participatory Extension events with diverse stakeholders to promote peer-to-peer learning, networking, and co-innovation. Two, five-hour, in-person sessions (n=27, n=31) with distinct participants were hosted in Pullman, WA, in February 2020. Using the LS toolkit, the agenda was storyboarded to facilitate a rapid co-innovation process progressing through the sharing of common purpose, identifying key questions and issues surrounding soil health, and preliminary solutions (Table 1).

The event was open invitation, announced throughout eastern Washington using emails targeting key stakeholder groups via listservs of over 1800 unique contacts from previous soil health activities. Stakeholders were identified and recruited to represent producers, crop advisors, policy interests, conservation agencies, industry, and academic research, within fruit and dryland wheat production systems (Figure 1). Demographic and motivation information was collected with registration to capture diversity among participant interests and soil health backgrounds. The approximate group size of 30 individuals was chosen to maximize opportunity to participate while maintaining connections between participants and facilitating group sharing. Initial seating was auditorium-style with rows of four to allow easy movement as 'structures' changed. Tables and writing materials were available around the perimeter of the room. All participants gave oral informed consent for data collection.

At the beginning of each event, facilitators presented ground rules that encouraged attentiveness and respect. Facilitators outlined activities, kept time using a bell, and guided reflection. The use of specific LS verbiage was limited. Each LS activity had a distinct and dynamic format, typically with a prompting question enabling participants to work through different stages of the innovation process in various group sizes, time limits, and space arrangements. Outcomes were assessed with a post-workshop evaluation form to determine the impact of the sessions on innovation processes and network building.

Table 1. Agenda of Storyboarded LS Activities to Lead Participants through Soil Health Innovation and Co-Production

Activity Objective	Liberating Structure Tool	Format
Rapidly share purpose of gathering and facilitate networking	Impromptu Networking	In groups of two, each person shares a two-minute response to group prompt
Make the purpose of the work together clear	Nine Whys	In pairs, each person answers repeated queries of 'Why is this important to you.' The uncovered underlying motivations are then shared with another pair.
Get practical and imaginative help from colleagues immediately	Troika Consulting	In groups of three, participants take turns playing the role of client with a question and consultants who offer advice on the question
Articulate the paradoxical challenges the group must confront to succeed	Wicked Questions	Alone, and then in a small group of 4-6 participants discuss and distill "How is it that X and Y can happen simultaneously?"
Discover, spark, and unleash local solutions to chronic problems	Discovery and Action Dialogues	In small groups participants discuss a series of seven progressive problem-solving questions
Rapidly generate and sift a group's most powerful actionable ideas	25/10 Crowd Source	Participants write their 'boldest' solution and ideas on notecards, notecards are rapidly exchanged, ranked, and shared with the group
Discover and focus on what each person has the freedom and resources to do now	15% solutions	Participants individually generate solutions that will generate largest impact with fewest resources, then form small groups to share and discuss

Note. Process adapted from Lipmanowicz and McCandless (2014)

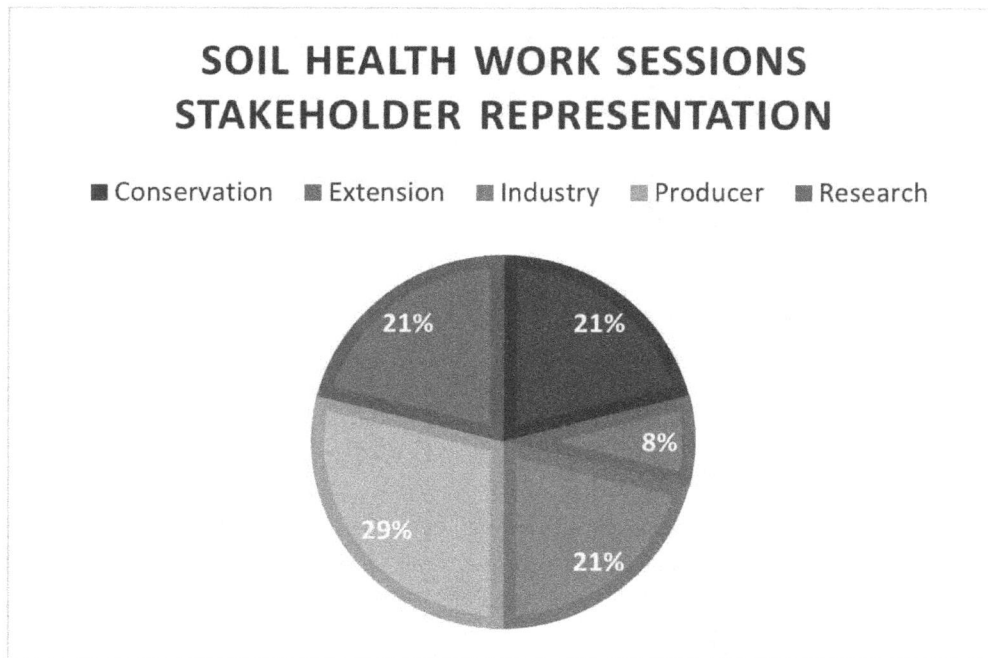

Figure 1. Stakeholder representation across both work sessions.

RESULTS AND DISCUSSION

The use of LS for participatory Extension was new to participants, who were primarily used to unidirectional content transfer. During the work sessions, participants identified three pathways forward to promote soil health in Washington State: policy and economic changes to allow producers more flexibility in managing soil health, consistent and effective soil health metrics and novel research on regenerative nutrient cycling processes, and an increase in community awareness of soil health. Adoption of a variety of soil management techniques (e.g., cover crops, effective microbial solutions, soil erosion prevention, etc.) with the intention of improving soil health on farms increases the subjective well-being of producers and their communities (Friedrichsen et al., 2021).

The sessions met the organizers' goals of enhancing peer-to-peer learning, building social networks, and fostering co-innovation with 80% of participants reporting increased knowledge, 80% reporting a new professional connection, and 92% indicating an intent to take action to increase soil health within the next six months (Table 2). The LS toolkit shifted participant experience to focus on co-knowledge production including increased sense of community, active learning, increased empathy of others' world views, creativity, and reflection (Table 3). These results indicate outcomes of expanded mental models, co-production of knowledge, and social learning (Jagannathan et al., 2020). These important shifts in participant experience are necessary to help support the transition from linear learning to a co-development of innovations paradigm.

To improve participant experience, we recommend considering hearing-impaired attendees, clearly stating the workshop goals at the beginning, and allowing participants to reflect on group goal achievement.

CONCLUSION

The novel use of Liberating Structures in agricultural Extension was effective at promoting co-innovation and building networks across stakeholder groups. LS tools were easy to learn and implement, resulting in positive experiences for facilitators and participants. Over 30 LS tools are freely available and can be applied towards a multitude of ends including road mapping and needs assessments, evaluation, brainstorming, problem solving, education, and almost any application where facilitators desire to engage every participant in the room.

Table 2. Evolution of Work Session

Participant self-reported response	% of survey respondents reporting
Expressed increased knowledge	80%
Reported a new professional connection	80%
Would recommend participation to a colleague	63%
Indicated taking action to increase soil health within the next 6 months	92%

Table 3. Participants' Perceived Impact of Workshop

Emergent Themes	Participant Quotes
Sense of community and social networking	"Knowing I'm not alone"
Active learning to foster co-development of Innovation	"It made us think right away. We had to put our thinking caps on early. As opposed to listening."
Development of empathy towards the perspective of other stakeholders' perceptions of soil health	"It gave me the opportunity to see how others in the agricultural industry think" "Helped understand issues farmers face—good discussion of issues and good networking"
Creative in-depth learning and reflection	"What exactly is [the] issue [of soil health]—helped me think this thru [sic]" "Think outside of box—our own personal box"

REFERENCES

Bell, S., & McAllister, J. (2021). *Sustainable agriculture through sustainable learning: An educator's guide to best practices for adult learning.* Technical Bulletin. Sustainable Agriculture Research and Education. https://www.sare.org/wp-content/uploads/Sustainable_Agriculture_Through_Sustainable_Learning.pdf

Bremer, S., & Meisch, S. (2017). Co-production in climate change research: Reviewing different perspectives. *WIREs Climate Change 8*(6), e482. doi.org/10.1002/wcc.482

Chumley, H., & Magrane, D. (2011). Energizing a faculty medical education retreat with liberating structures. *Medical Education 45*(5), 508-535. https://ur.booksc.org/book/9291736/3b568b

Friedrichsen, C. N., Hagen-Zakarison, S., Friesen, M. L., McFarland, C. R., Tao, H., & Wulfhorst. J. D. (2021). Soil health and well-being: Redefining soil health based upon a plurality of values. *Soil Security 2.* https://doi.org/10.1016/j.soisec.2021.100004

Jagannathan, K., Arnott, J. C., Wyborn, C., Klenk, N., Mach, K. J., Moss, R. H., Sjostrom, K. D. (2020). Great expectations? Reconciling the aspiration, outcome, and possibility of co-production. *Current Opinion in Environmental Sustainability 42*, 22–29. doi.org/10.1016/j.cosust.2019.11.010

Janzen, H. H., Janzen, D. W., & Gregorich, E. G. (2021). The 'soil health' metaphor: Illuminating or illusory? *Soil Biology and Biochemistry 159*, 108167. doi.org/10.1016/j.soilbio.2021.108167

Klerkx L., Schut, M., Leeuwis, C., Kilelu, C. (2012). Advances in knowledge brokering in the agricultural sector: Towards innovation system facilitation. IDS Bulletin 43(5), 53-60. Doi.org/10.1111/j.1759-5436.2012.00363.x

Lipmanowicz, H. & McCandless, K. (2014). *The surprising power of liberating structures: Simple rules to unleash a culture of innovation.* Liberating Structures Press.

Mahoney, J. S., Lewin, L., Beeber, L., & Willis, D. G. (2016). Using liberating structures to increase engagement in identifying priorities for the APNA research council. *Journal of the American Psychiatric Nurses Association. 22*(6), 504-507. https://doi.org/10.1177/1078390316663308

Mallete, C., & Rykert, L. (2018). Promoting positive change in nursing faculties: Getting to maybe through liberating structures. *Journal of Professional Nursing 34*(3) 161-166. Doi.org/10.1016/j.profnurs.2017.08.001

Rogers, E. M. (2003). *Diffusion of innovations* (5th edition). Free Press.

Singhal, A., Perez, L. E., Stevik, K., Monness, E., & Svenkerud, P.J. (2020). Liberating structures as pedagogical innovation for inclusive learning: A pilot study in a Norwegian university. *Journal of Creative Communications 15*(1), 35-52. https://doi.org/10.1177/0973258619875600

Torbert, W. (1991). *The power of balance: Transforming self, society, and scientific inquiry.* Sage Publications.

Wang, T., Jin, H., Kashu, B. B., Jacquet, J. B., & Kumar, S. (2019). Soil conservation practice adoption in the northern Great Plains: Economic versus stewardship motivations. *Journal of Agricultural and Resource Economics 44*(2), 404-421. doi.org/10.22004/ag.econ.287989

JOURNAL OF
Extension

Ideas at Work

Volume 60, Issue 2, 2022

Transformative Partnerships: Expanding Extension's Capacity to Support Texans with Developmental Disabilities

ANDREW B. CROCKER[1], MORGAN D. BRADLEY[1], SHELBY D. VAUGHN[1], AND BETH STALVEY[2]

AUTHORS: [1]Texas A&M AgriLife Extension Service. [2]Texas Council for Developmental Disabilities.

Abstract. New partnerships to reach new audiences are key to Extension's future (Harder, 2019). But partnership is enhanced through shared decision-making, co-creation of content, and leveraging non-overlapping expertise and experience (Bertsch et al., 2020; Israilov & Cho, 2017; Ostrom, 1996). Texas A&M AgriLife Extension Service partnering with the Texas Council for Developmental Disabilities provides a novel approach to using statewide presence to the benefit of a partner seeking to expand its footprint (Alford, 2014; Ostrom, 1996) and is, itself, an outcome (Voorberg et al., 2015). Additionally, Texans with disabilities benefit through greater access to the education and resources the partnership produces.

INTRODUCTION

While the Cooperative Extension System nationally has engaged in serving clientele with developmental disabilities, most efforts have been program-specific and heavily youth-focused rather than more broadly applicable to Extension's work (Taylor-Winney et al., 2019). Brill (2011), Angima et al. (2016), and Keywood and Brill (2020) discuss strategies for more inclusive, accessible Extension education; Peterson et al. (2012) explores the Extension professionals' experience with inclusive programming and their trainings needs. However, the literature does not provide examples of capacity building at the local level to ensure Extension personnel and programming are reaching individuals with intellectual or developmental disabilities.

More than 3.2 million community-dwelling Texans have a disability (U.S. Census, 2019). With an estimated 52% of individuals with disability living in just 10 of Texas' 254 counties (Texas Workforce Investment Council, 2019), experiences with support, access to information, and resources vary depending on where an individual resides as well as the type of disability. The mission of the Texas Council for Developmental Disabilities (TCDD) is to create change so that all people with disabilities are fully included in their communities and exercise control over their own lives. TCDD advances this mission by funding innovative grant projects across the state, training to empower Texans to advocate for the change they wish to see, sharing resources, and organizing educational programs (Administration for Community Living [ACL], 2021; TCDD, n.d.-a). TCDD is governed in accordance with the Developmental Disabilities Assistance and Bill of Rights Act and reports to a 27-member Council consisting of individuals with developmental disabilities and family members, appointed by the Texas Governor and representatives from state agencies that provide services and support. TCDD is part of a national network of developmental disability councils that exist in every state and territory in the United States (ACL, 2021).

COMMUNITY OUTREACH COORDINATOR PROJECT

In 2019, Texas A&M AgriLife Extension Service (AgriLife Extension) entered an innovative, multi-year, collaborative partnership with TCDD to provide education, training, and resources for individuals with intellectual or developmental disabilities, family supports, partners, and providers, including Extension professionals. Through the presence of regional Community Outreach Coordinators (Coordinators) housed at AgriLife Extension facilities, TCDD gains the visibility, benefit, and reputation of being active and engaged throughout Texas, not just in its population centers (Figure 1). In addition to physical support, TCDD benefits from the Cooperative Extension know-how in approaches to outreach and education (Franz & Townson, 2008).

AgriLife Extension benefits through the recognition of its value as a presence and thought leader in local communities, willing to engage in collaborative public scholarship with a public sector partner (Franz, 2003, 2011; Harder, 2019; Kalambokidis, 2004) on behalf of a constituency craving information and supports. The project also seeks to achieve some of the professional development recommendations identified by Peterson et al. (2012) and Keywood and Brill (2020) including, but not limited to, inclusive programming, expansion of disability resources, and increased disability awareness.

INNOVATIVE EDUCATIONAL APPROACHES

The Community Outreach Coordinator Project brings a unique focus to Extension education through disability-related topic areas and broadening existing efforts related to diverse, inclusive, and equitable audiences and programming. Franz and Townson (2008) describe the content spectrum as information sharing and the process spectrum as educational program delivery (see Table 1). During the first phase of the partnership, AgriLife Extension and TCDD focused on low content activities through *service* and *facilitation* approaches. Service activities included connecting with regional networks, community coalitions, and committees and increasing stakeholder engagement of disability-focused and non-disability-focused organizations; facilitation activities included the launch of community listening sessions across the state to gather information on regional disability-related issues and to inform future activities.

As the partnership continues to evolve, AgriLife Extension and TCDD are focusing on high content activities through content transmission and transformative education approaches. Content transmission activities include dissemination of disability-related resources and education through bimonthly newsletters, localized resource guides, and ongoing webinar series; transformative learning activities include opportunities that allow/will allow disability experts to provide single-event training opportunities for county Extension agents, childcare providers, and other service professionals. Education delivered within this approach is focused on applicable, practical training that clientele can adopt in their everyday practice.

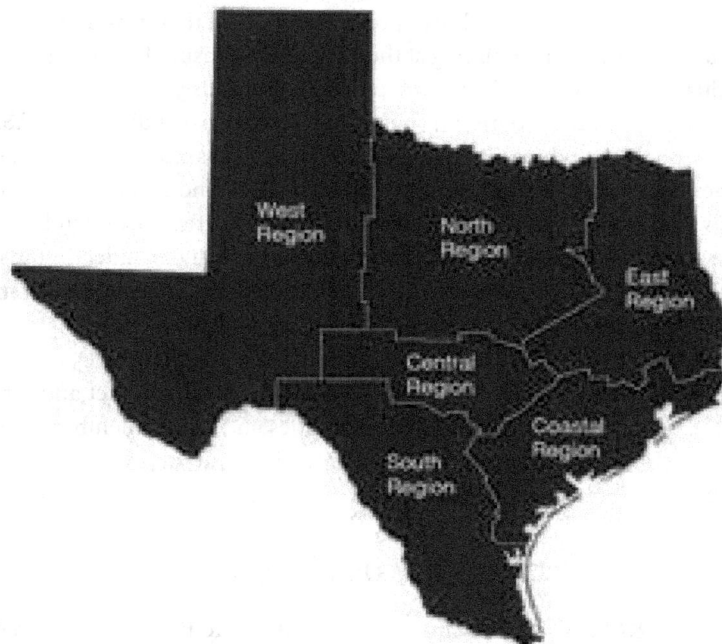

Figure 1. Community Outreach Coordinator regions. Reprinted from *Connect with Your Regional Coordinator* (TCDD, n.d.-b). Copyright 2021 by TCDD.

Table 1. Community Outreach Coordinator Project Transformative Partnership Model

		CONTENT	
		Low	**High**

PROCESS	**High**	*Facilitation*
		• Regional listening sessions
		• Surveys related to barriers to
		  • community engagement
		  • impact of COVID-19 pandemic on community programming
			Transformative Education
			• Educational series/trainings related to
			  • employing persons with disabilities
			  • emergency management
			  • worksheet for inclusive Extension programming
			  • inclusive childcare

	Low	*Service*
		• Committees
		• Community coalitions
		• Stakeholder engagement
		• Developing/growing new/existing community partnerships
			Content Transmission
			• Newsletters
			• Community resource guides
			• Educational presentations related to
			  • aging with disability
			  • special education
			  • healthcare transitions

Note. This table is adapted from the Extension Educational Approaches Model (Franz & Townson, 2008).

CO-CREATION AS A MEANS FOR TRANSFORMATIVE EDUCATION

Transformative education is defined as changing pre-existing knowledge through newly acquired learning material, critical discourse, and self-reflection to better assimilate to an everchanging environment (Mezirow, 1991). Neither AgriLife Extension nor the Cooperative Extension System as a whole can achieve the high process, high content educational approach of transformative education in an area like developmental disability without "challeng[ing] the traditional expert model and the dominant mind-set of many Extension professionals" (Bertsch et al., 2020) and engaging in co-creation of educational products with public and private sector partners. Franz (2003) describes a positive linear relationship between increased Extension partnerships with community networks and increased adaptions of programmatic changes. This sentiment is echoed in the literature supporting public sector co-creation (Alford, 2014; Israilov & Cho, 2017; Ostrom, 1996; Voorberg et al., 2015). Through co-creation, AgriLife Extension is still an expert in a successful outreach, education, and community capacity building model that has worked for more than a century, and community partners are hungry for access to that expertise and experience to better serve their constituencies.

CONCLUSION

The partnership between Texas A&M AgriLife Extension Service and the Texas Council for Developmental Disabilities has evolved to a mutually beneficial level where both parties continue to engage in co-creation of educational products, build community capacity, and prove a replicable model for other states and with other disciplines. Partnership with agencies and organizations with whom Extension does not compete for the same audience, dollar, or expertise provides an opportunity to revitalize its relevance in communities and with clientele.

REFERENCES

Administration for Community Living. (2021). *State councils on developmental disabilities.* https://acl.gov/programs/aging-and-disability-networks/state-councils-developmental-disabilities

Alford, J. (2014). The multiple facets of co-production: Building on the work of Elinor Ostrom. *Public Management Review, 16*(3), 299–316. https://doi.org/10.1080/14719037.2013.806578

Angima, S., Etuk, L., & Maddy, D. (2016). Accommodating Extension clients who face language, vision, or hearing challenges. *Journal of Extension, 54*(4). http://archives.joe.org/joe/2016august/a3.php

Bertsch, B., Heemstra, J., & Golick, D. (2020). Preparing to cocreate: Using learning circles to ready Extension professionals for meaningful stakeholder engagement. *Journal of Extension, 58*(3). https://archives.joe.org/joe/2020june/a1.php

Brill, M. F. (2011). Teaching the special needs learner: When words are not enough. *Journal of Extension, 49*(5). http://archives.joe.org/joe/2011october/tt4.php

Franz, N. K. (2003). Transformative learning in Extension staff partnerships: Facilitating personal, joint, and organizational change. *Journal of Extension, 41*(2). http://archives.joe.org/joe/2003april/a1.php

Franz, N. K. (2011). Advancing the public value movement: Sustaining Extension during tough times. *Journal of Extension, 49*(2). http://archives.joe.org/joe/2011april/comm2.php

Franz, N. K. & Townson, L. (2008). The nature of complex organizations: The case of Cooperative Extension. In M. Braverman, M. Engle, M. Arnold, & R. Rennekamp (Eds.), *Program evaluation in a complex organizational system: Lessons from Cooperative Extension. New Directions for Evaluation, 120* (pp. 5–14). Jossey-Bass.

Harder, A. (2019). Public value and partnerships: Critical components of Extension's future. *Journal of Extension, 57*(3). https://archives.joe.org/joe/2019june/comm1.php

Israilov, S. & Cho, H. J. (2017). How co-creation helped address hierarchy, overwhelmed patients, and conflicts of interest in health care quality and safety. *AMA Journal of Ethics, 19*(11), 1139–1145. https://doi.org/10.1001/journalofethics.2017.19.11.mhst1-1711

Kalambokidis, L. (2004). Identifying the public value in Extension programs. *Journal of Extension, 42*(2). https://archives.joe.org/joe/2004april/a1.php

Keywood, J. R. & Brill, M. F. (2020). Developmental disabilities training series. *Journal of Extension, 58*(3). http://archives.joe.org/joe/2020june/tt3.php

Mezirow, J. (1991). *Transformative dimensions of adult learning.* Jossey-Bass.

Ostrom, E. (1996). Crossing the great divide: Coproduction, synergy, and development. *World Development, 24*(6), 1073–1087. https://doi.org/10.1016/0305-750x(96)00023-x

Peterson, R. L., Grenwelge, C., Benz, M. R., Zhang, D., Resch, J. A., Mireles, G. & Mahadevan, L. (2012). Serving clients with disabilities: An assessment of Texas FCS Agents' needs for implementing inclusive programs. *Journal of Extension, 50*(6). http://archives.joe.org/joe/2012december/a7.php

Taylor-Winney, J., Xue, C., McNab, E., & Krahn, G. (2019). Inclusion of youths with disabilities in 4-H: A scoping literature review. *Journal of Extension, 57*(3). http://archives.joe.org/joe/2019june/a1.php

Texas Council for Developmental Disabilities. (n.d.-a). *Legal authorization.* Retrieved March 31, 2021, from https://tcdd.texas.gov/about/legal-authorization/

Texas Council for Developmental Disabilities. (n.d.-b). *Connect with your regional coordinator.* Retrieved on February 28, 2022, from https://tcdd.texas.gov/stay-connected/regional-coordinators/Texas Workforce Investment Council. (2019). *People with disabilities: A Texas profile.* https://gov.texas.gov/uploads/files/organization/twic/People-With-Disabilities-2019.pdf

U.S. Census Bureau. (2019). *2015–2019 American Community Survey 5-year estimates: Disability characteristics–Texas.* Retrieved March 16, 2021, from https://data.census.gov/cedsci/table?q=disability&g=0400000US48&tid=ACSST5Y2019.S1810&hidePreview=true

Voorberg, W. H., Bekkers, V. J. J. M., & Tummers, L. G. (2015). A systematic review of co-creation and co-production: Embarking on the social innovation journey. *Public Management Review, 17*(9), 1333–1357. https://doi.org/10.1080/14719037.2014.930505

JOURNAL OF Extension

Ideas at Work

Volume 60, Issue 2, 2022

Interdisciplinary Team Addresses Cotton Leafroll Dwarf Virus in Alabama

Kassie N. Conner[1], Edward Sikora[1], Jenny Koebernick[1], and Marcio Zaccaron[1]

AUTHORS: [1]Auburn University.

Abstract. A multi-state and interdisciplinary team was formed to address the Extension and research needs of CLRDV, an emerging cotton disease with high potential impact for U.S. cotton production. In 2017, CLRDV was identified in AL and Auburn University immediately formed an interdisciplinary working group composed of plant breeders, plant pathologists, entomologists, and agronomists. Since then, scientists from ten other states have joined the CLRDV group. Thus, allowing research to be coordinated efficiently and best deploy limited resources to attend the stakeholder's needs. The CLRDV group produces and shares new and relevant information with the scientific community and cotton producers alike.

INTRODUCTION

Cotton (*Gossypium hirsutum*) is one of the most economically important crops in the United States. During 2017, *Cotton leafroll dwarf virus* (CLRDV) (genus *Polerovirus*, family *Luteoviridae*) was first identified in the United States in cotton on approximately 50,585 ha from Alabama (Avelar & Schrimsher et al., 2019). CLRDV is transmitted by the cotton aphid (*Aphis gossypii*). The virus has since been found across the cotton belt in North Carolina, South Carolina, Georgia, Tennessee, Florida, Mississippi, Louisiana, Texas, Arkansas, Oklahoma, and Kansas (Aboughanem-Sabanadzovic et al., 2019; Alabi et al., 2019; Ali & Mokhrari, 2020; Ali et al., 2020; Faske et al., 2020; Iriarte et al., 2020; Price et al., 2020; Tabassum et al., 2019; Thiessen et. al., 2020; Wang et al., 2020). Symptoms of CLRDV consist of: stunting due to internodal shortening; leaf rolling; leaf cupping; leaf petiole and vein reddening; distorted new growth; rugosity; wilting of plants; reduced flower and boll size; accentuated verticality; and sterility. Symptoms can vary based on: plant age at time of infection, cotton variety, environmental conditions at or after infection, and nutritional status of the plant/field. In 2017, disease incidence based on visual symptoms (i.e., leaf crinkle) ranged from 3 to 30%, and yield loss was estimated at an average of 560 kg/ha which was valued at $19 million dollars (Avelar, & Schrimsher, et al., 2019).

Whole genome sequences derived from symptomatic leaf samples collected during 2018 showed that the strain of CLRDV found in the United States is different from the "typical" and "atypical" strains previously found in South America. This confirmed that the CLRDV strain detected in in the United States is unique (Avelar, Sobrinho, et al., 2019). Whole genome sequences of CLRDV-AL isolates collected from other states across the cotton belt are similar, which suggests a single introduction of the virus into the United States (Aboughanem et al., 2021).

DEVELOPMENT OF CLRDV WORKING GROUP

The development of resistant cotton cultivars will be the best defense against CLRDV in the United States. Seed companies have germplasm with resistance to "typical" and "atypical" CLRDV strains, but these have proven susceptible to the CLRDV-AL found in the United States. In 2018, Alabama Extension Specialists and Auburn University research faculty from multiple disciplines (Plant Breeding, Agronomy, Entomology, and Plant Pathology) formed a CLRDV working group. The group has focused efforts on Extension outreach and research projects to address this emerging disease. Initially, a statewide survey was conducted to determine the distribution of the pathogen in Alabama. At this time, CLRDV had been found in 43 of the state's 67 counties, including all major cotton production areas of Alabama. From 2018 to 2019, a weed survey was conducted to identify alternate hosts of the virus, and multiple species were identified as natural hosts (K. Conner, personal communication).

In 2019, a multi-state CLRDV Sentinel Plot Working Group was formed that included Extension Specialists and researchers from 11 cotton belt states (Alabama, Virginia, North Carolina, South Carolina, Georgia, Florida, Mississippi, Arkansas, Louisiana, Tennessee, and Texas). This multi-state group meets regularly to share updates on Extension and research efforts on CLRDV. The CLRDV sentinel plots, which included 15 sites across 10 states (with five regional sites in AL), were first established in 2019 to assess planting dates, environmental factors, varietal selection, and yield impacts of CLRDV. Leaf samples were collected twice per season by Auburn University personnel from all sentinel locations and tested for the presence of CLRDV at the Auburn University Plant Diagnostic Lab. Results from the sentinel plot system increased awareness of the disease to growers and agribusiness clientele and allowed the industry to clearly see the distribution of the disease in the United States.

Extension/research efforts at Auburn have included identifying the aphid vector and characterizing transmission efficiency (Heilsnis et al., 2020). A robust breeding program at Auburn has allowed researchers to evaluate over 2,000 cotton breeding lines to screen for resistance to the disease. A leaf disc assay was also developed to improve the efficiency of germplasm screening (Heilsnis et al., 2021). Members of the Auburn CLRDV working group have set up insecticide trials to determine the efficacy of insecticide treatments and found that intensive insecticide programs do not result in a reduction of virus spread (A. Jacobson, personal communication). The Auburn Plant Diagnostic Lab tested over 4,000 cotton samples in 2019, and 5,200 samples in 2020 for the presence of CLRDV.

CONCLUSION AND RECOMMENDATIONS

Traditional Extension methods were utilized to inform growers about CLRDV including presentations at county and regional production meetings (over 35 presentations from 2018 to 2020), farm visits and field days, pest alerts issued in news articles, blogs and the ACES website, and updates on CLRDV through Twitter and Facebook (Sikora et al., 2009). In-service trainings were conducted annually for Regional Extension Agents and members of the ACES Agronomic Crop Team to provide updates on CLRDV in Alabama. A CLRDV diagnostic field guide was developed for growers to assist in disease scouting and includes management recommendations for the disease (Conner et al., 2021). To date, the work from this interdisciplinary team of Extension and research personnel has led to 12 peer reviewed journal articles and over $25 million in funding.

Current management recommendations for growers include: 1) Cotton in areas at high risk for infection should be planted early. Disease incidence and subsequent symptom severity is much greater in late-planted cotton. 2) Many weed species have been found to be natural hosts of CLRDV and may have a role in overwintering of the pathogen and likely act as a source of primary inoculum. An aggressive weed control program around field borders and nearby ditch banks prior to planting may be helpful in delaying virus movement from overwintering hosts into cotton fields. 3) Destroy cotton stalks following harvest as well as kill ratoon and volunteer cotton with herbicides or tillage equipment. They may serve as a bridge between an overwintering host and the newly planted cotton crop. 4) Long term establishment of resistant cotton cultivars will be the best defense against CLRDV.

REFERENCES

Aboughanem-Sabanadzovic, N., Allen, T. W., Wilkerson, T. H., Conner, K. N., Sikora, E. J., Nichols, R. L., & Sabanadzovic, S. (2019). First report of cotton leafroll dwarf virus in upland cotton (*Gossypium hirsutum* L.) in Mississippi. *Plant Disease*, 103(7), 1798. https://doi.org/10.1094/PDIS-01-19-0017-PDN

Aboughanem, N., Allen, T. W., Wilkerson, T. H., Scheffler, J., & Sabanadzovic, S. (2021). Study of cotton leafroll dwarf virus in Mississippi: State of the art (2020) [Conference presentation]. Beltwide Cotton Conferences, virtual.

Alabi, O. J., Isakeit, T., Vaughn, R., Stelly, D., Conner, K., Gaytan, B., Villegas, C., Hitzelberger, C., De Santiago, L., Monclova-Santana, C., & Brown, J. (2020). First report of cotton leafroll dwarf virus infecting upland cotton (*Gossypium hirsutum* L.) in Texas. *Plant Disease*, 104(3), 998. https://doi.org/10.1094/PDIS-09-19-2008-PDN

Ali, A., & Mokhtari, S. (2020). First report of cotton leafroll dwarf virus infecting cotton (*Gossypium hirustum*) in Kansas. *Plant Disease, 104*(6),1880. https://doi.org/10.1094/PDIS-12-19-2589-PDN

Ali, A., Mokhtari, S., & Ferguson, C. (2020). First Report of Cotton Leafroll Dwarf Virus from Cotton (*Gossypium hirsutum*) in Oklahoma. *Plant Disease, 104*(9), 2531. https://doi.org/10.1094/PDIS-03-20-0479-PDN

Avelar, S., Schrimsher, D. W., Lawrence, K. S., & Brown, J. K. (2019). First report of cotton leafroll dwarf virus associated with cotton blue disease in Alabama. *Plant Disease, 103*(3), 592–592. https://doi.org/10.1094/PDIS-09-18-1550-PDN

Avelar, S., Sobrinho, Roberto R., Conner, Kassie, Nichols, Robert L., Lawrence, Kathy S., & Brown, Judith K. (2019). Characterization of the complete genome and P0 protein for a previously unreported genotype of cotton leafroll dwarf virus, an introduced polerovirus in the USA. *Plant Disease, 104*(3), 780–786. https://doi.org/10.1094/PDIS-06-19-1316-RE

Faske, T. R., Stainton, D., Aboughanem-Sabanadzovic, N., & Allen, T. W. (2020). First report of cotton leafroll dwarf virus from upland cotton (*Gossypium hirsutum*) in Arkansas. *Plant Disease, 104*(10), 2742. https://doi.org/10.1094/PDIS-12-19-2610-PDN

Conner, K. N., Sherer, A., Hagan, A., Koebernick, J., Jacobson, A., Bowen, K. L., Sikora, E. J., Graham, S., & Brown, S. (2021). Cotton leafroll dwarf virus. ANR-2539. https://www.aces.edu/blog/topics/crop-production/cotton-leafroll-dwarf-virus/

Heilsnis, B., Conner, K., Koebernick, J., & Jacobson, A. (2020). *Transmission of cotton leafroll dwarf virus by Aphis gossypii* [Conference presentation]. Beltwide Cotton Conferences, Austin, TX, United States.

Heilsnis, B., Koebernick, J., Conner, K., & Jacobson, A. (2021). Rapid screening for resistance to CLRDV: Development of an aphid assay [Conference presentation]. Beltwide Cotton Conferences, virtual.

Iriarte, F., Dey, K. K., Small, I. M., Conner, K., O'Brien, K., Johnson, L., Savery, C., Carter, E., Sprague, D., Wright, D. L., Nichols, R. L., Mulvaney, M., & Paret, M. L. (2020). First report of cotton leafroll dwarf virus (CLRDV) in Florida. *Plant Disease, 104*(10), 2744. https://doi.org/10.1094/PDIS-10-19-2150-PDN

Price, T., Valverde, R., Singh, R., Davis, J., Brown, S., & Jones, H. (2020). First report of cotton leafroll dwarf virus in Louisiana. *Plant Health Progress, 21*(2), 142–143. https://doi.org/10.1094/PHP-03-20-0019-BR

Sikora, E. J., Delaney, D., & Delaney, M. A. (2009). Developing an innovative team approach to address a newly introduced disease of soybeans in the United States. *Journal of Extension, 47*(4). https://archives.joe.org/joe/2009august/iw7.php

Tabassum, A., Bag, S., Roberts, P., Suassuna, N., Chee, P., Whitaker, J. R., Conner, K. N., Brown, J., Nichols, R. L., & Kemerait, R. C. (2019). First report of cotton leafroll dwarf virus infecting cotton in Georgia, USA. *Plant Disease, 103*(7), 1803. https://apsjournals.apsnet.org/doi/pdf/10.1094/PDIS-12-18-2197-PDN

Thiessen, L., Schappe, T. L., Zaccaron, M., Conner, K., Koebernick, J., Jacobson, A., & Huseth, A. (2020). First report of cotton leafroll dwarf disease caused by cotton leafroll dwarf virus affecting cotton in North Carolina. *Plant Disease, 104*(12), 3275. https://doi.org/10.1094/PDIS-02-20-0335-PDN

Wang, H., Greene, J., Mueller, J. D., Conner, K., & Jacobson. A. (2020). First report of cotton leafroll dwarf virus in cotton fields of South Carolina. *Plant Disease, 104*(9), 2532. https://doi.org/10.1094/PDIS-03-20-0635-PDN

Unique Conference Design Showcases Small Towns, Highlights Entrepreneurs, and Strengthens Capacity

CAREY ANDREW NORTHROP[1], KATHERINE M. JAMIESON[1],
PARKER B. JONES[1], MARY A. REILLY[1], AND TYLER AUGST[1]

AUTHORS: [1]Michigan State University Extension.

Abstract. Michigan State University Extension (MSUE)'s annual conference, Connecting Entrepreneurial Communities (CEC), has served as a catalyst for entrepreneurial ecosystems across Michigan since 2012. Designed by MSUE for small towns, CEC has gained national interest as evidenced by the adoption of this conference model by four other Extension services. This article outlines the unique conference design, details the partnership between Extension and host communities, and explores conference evaluation data validating the need to continue this programming. Lessons learned and successes to date are provided to ensure readers learn the value this unique conference format has in Extension entrepreneurship programming nationally.

INTRODUCTION

This article highlights a place-based conference model developed by Michigan State University Extension (MSUE) known as Connecting Entrepreneurial Communities (CEC). CEC is an unconventional conference embedded in a small town with the goal of showcasing the host community, highlighting entrepreneurs, and strengthening organizational capacity. Designed to address several elements known to support small town entrepreneurial ecosystems, CEC focuses on networking, policy, human capital, culture, and support (Isenberg, 2010; Roundy, 2017). Additionally, conferences and other types of events can serve as the genesis for entrepreneurial networks (Case & Harris, 2012; cited in Roundy, 2017, p. 247). After describing the conference's unique design, we explain the roles of the host community, external partnerships, and evaluations. This unique conference design could be widely applied in Extension entrepreneurship programming.

CONFERENCE DESIGN

Launched by MSUE in 2012, CEC is *a conference for small towns held in a small town*. Unique in multiple ways, CEC works in partnership with a different host community annually and takes place from noon to noon over two days, typically in early October. Host communities are walkable small towns and range in population size between 2,500 to 10,000 people. Unlike traditional conferences based in a hotel or conference center, CEC embeds over 20 educational sessions within the businesses of a host community's downtown. For example, coffee shops, art galleries, breweries/restaurants, and museums, to name a few, have served as settings for educational sessions. The CEC schedule intentionally weaves 30-minute breaks between sessions so participants can network and patronize local businesses at a leisurely pace (see Appendix A).

CEC is designed to highlight success stories of local entrepreneurs within the host community in multiple ways. First, at the beginning of an educational session, local business owners are invited to briefly share a synopsis of how they began their business. Second, two local entrepreneurs are featured as plenary and closing keynote speakers. The keynote speakers are generally selected because their business has been rooted in the community over generations or because they are an entrepreneur with more recent influence, but they both have a strong connection to their local community. First-hand entrepreneurial stories relay both success and failure and portray the "encouraging mindset and spirit of facilitation" critical to entrepreneurial ecosystems (Russell, 2006; cited in Butler, 2006, p. 112).

Finally, to strengthen a community's organizational capacity, CEC's promotional materials encourage communities to create a team to attend the conference. Teams typically range in size from three to five professionals and represent organizations charged with supporting local businesses and entrepreneurs, such as a chamber of commerce. Although a team is not required, this strategy can strengthen the attending team's networks and resources and position them to host CEC in the future. Due to the unique design of CEC, prospective host communities are required to have attended at least one previous CEC event as a team. This practice ensures hosting communities can experience the unique conference design and unconventional format before taking on the role of a CEC host community.

MSUE AND HOST COMMUNITY ROLES

Partnership with the host community is an intentional part of the conference design and comes with pivotal responsibilities for both the host community and MSUE. Host community teams identify and secure a) local entrepreneur(s) to serve as keynote speaker(s), b) locations for educational sessions within local businesses and organizations, c) local sponsorships, and d) marketing materials to showcase the community to participants. The host community plays an integral role in selecting session venues based on interest, access, auditory challenges, and seating capacity for upwards of 20 participants. Ideally, the venue pairs with the session theme, such as an art gallery hosting an educational session on art as a community economic driver.

Within MSUE, programming and planning for CEC draws from several multidisciplinary teams. MSUE's CEC planning team is responsible for a) selecting a host community, b) co-planning the event with the host community, c) securing educational session speakers from MSUE, d) managing registration, and e) ensuring the host community executes their responsibilities. To ensure both the MSUE team and the host community teams are in unison throughout the typical eight-month planning period, a liaison from each team serves both groups. The liaison position was created to address previous complications experienced in the early years of CEC.

Early in CEC's tenure, MSUE failed to provide clear expectations to the host community team, resulting in missed deadlines for selecting venues and keynote speakers. When CEC was nearly canceled in that instance, the MSUE team formalized host community protocols. Established protocols actively guide the partnership by clearly articulating expectations to host communities in advance, requiring that the community send a team to a CEC conference prior to being considered as a host community, identifying a liaison to serve on both the MSUE and host community teams, and requiring the host community secure a predetermined amount of the conference costs in sponsorships to create greater ownership.

The MSUE and host community teams actively seek partnerships with entities such as state agencies and/or nonprofit organizations to support the conference. For example, in 2019 Michigan Economic Development Corporation (MEDC) co-sponsored CEC for the first time. MEDC staff led educational sessions and provided targeted marketing to Michigan's Main Street communities. The resources from MEDC and other partner organizations frequently come in the form of educational content expertise, conference sponsorship dollars, and state and national promotion.

EVALUATION AND METHODS

The conference attendees receive an evaluation at the end of the conference to assess learning and gauge intent to apply the information learned at CEC (see Appendix B). Each attendee receives a paper evaluation during the closing keynote speaker session. Organizers encourage completion of the evaluation during closing remarks, and MSUE staff collect evaluations at the exits. A survey link with the evaluation is sent out a week after the conference to give those who were not able to complete a paper survey another opportunity to complete the evaluation. Between 2012 and 2019, the overall evaluation response survey rate for CEC was 42%.

As of 2019, ten Michigan communities have hosted the CEC conference. Since its start in 2012, 1,006 individuals have participated in the CEC conference representing Michigan's 83 counties,12 other states, and two countries. Over this period, participants overwhelmingly confirmed (98.7%, n = 306) that they liked the approach of holding sessions in downtown businesses. A majority (90.8%, n = 351) of the participants reported that they planned to create or expand community-based activities or initiatives to support local business and entrepreneurs.

The MSUE planning team made minor refinements to the evaluation tool in 2019 to remove items not yielding new insights and to reframe other items to assess intended application of knowledge gained. At the 2019 conference, the most recent in-person event, 27% (n = 27) of participants completed the evaluation. Of these respondents, 100% agreed or strongly agreed that "educational sessions were engaging" and the "content presented is replicable in my community." As part of the evaluation, participants were also presented with a list of potential actions to take as a result of their participation in CEC such as "integrate tourism and natural resources into economic strategy" and "provide youth entrepreneurship programming" (see Appendix B). Attendees from the 2019 CEC indicated a total of 80 anticipated actions on the evaluation. As an example, the city of Alma, Michigan learned about a library hot-spot rental program during an educational session at CEC 2019. They applied this knowledge to create their own program just ahead of the COVID-19 pandemic and aided students learning from home who would not otherwise have had internet access (K. Phillips, personal communication, April 20, 2020).

CONCLUSION

In summary, CEC's unique conference design is highly engaging and supports participants' ability to replicate ideas and knowledge learned at the conference. The partnership between the host community and MSUE plays a fundamental role in the success of CEC. Conference planning protocols designed by MSUE ensure that small town hosts and local entrepreneurs are highlighted while ensuring that participating individuals and teams receive content that strengthens their capacities to create strong entrepreneurial ecosystems. MSUE's CEC model has garnered national attention within Extension and, as a result, has been replicated in four other states (Table 1). As of 2021, universities in three additional states are in the planning stages of delivering their first CEC conference. MSUE's CEC model should have a place within Extension programming nationally as a collaborative approach to strengthening small town entrepreneurial ecosystems.

Table 1. Other States Engaging the CEC Model

State	Year Launched CEC Conference	Planning for CEC Conference
University of Nebraska	2017	
South Dakota State University	2018	
North Dakota State University	2019	
University of Minnesota	2019	
University of Missouri		x
Pennsylvania State University		x
University of New Hampshire		x

REFERENCES

Case, S., & Harris, D. (2012) *The startup uprising: Eighteen months of the Startup America Partnership.* Ewing Marion Kauffman Foundation.

Isenberg, D. J. (2010). How to Start an Entrepreneurial Revolution. *Harvard Business Review, 88*(6), pp. 41–50).

Roundy, P. (2017) "Small town" entrepreneurial ecosystems: Implications for developed and emerging economies. *Journal of Entrepreneurship in Emerging Economies, 9*(3), 244.

Russell, R. (2006). The contribution of entrepreneurship theory to the TALC model. In R. W. Butler (Ed.), *The tourism area life cycle, Vol. 2: Conceptual and theoretical issues* (pp. 105–123). Channel View Publications.

APPENDIX A. SAMPLE AGENDA FOR CEC CONFERENCE

Day 1

11:00 am	Registration/check-in with lunch and networking
12:00–1:00 pm	Opening session and keynote speaker
1:00–1:30	Travel to breakout sessions located in various businesses downtown
1:30–2:15	Breakout session 1
2:15–2:45	Travel to breakout sessions
2:45–3:30	Breakout session 2
3:30–4:00	Travel to breakout sessions
4:00–4:45	Breakout session 3
4:45–5:30	Break
5:30–7:30	Reception and networking event (heavy appetizers). Participants are encouraged to have dinner elsewhere and continue networking.

Day 2

7:30–8:30 am	Continental breakfast
8:30–9:00	Warm-up, wake-up, idea sharing! 30 ideas in 30 minutes
9:00–9:15	Travel to breakout session
9:15–10:00	Breakout session 4
10:00–10:15	Travel to breakout session
10:15–11:00	Breakout session 5
11:00–11:15	Travel to keynote location
11:15–12:00 pm	Keynote speaker and wrap-up
12:00	Evaluation and adjourn

APPENDIX B. SAMPLE EVALUATION FOR CEC CONFERENCE

Which of the following do you plan to do as a result of your participation in this program? (Check all that apply.)

- Seek out more knowledge on economic and workforce trends.
- Provide youth entrepreneurship programming and resources.
- Apply knowledge on agriculture, arts, and/or cultural development towards economic development. strategies.
- Integrate tourism and sustainable natural resource use into economic strategy.
- Make use of new and/or existing regional and community assets.
- Apply talent attraction and retention strategies toward community and economic prosperity.

	Strongly Agree	Agree	Disagree	Strongly Disagree
The educational sessions were engaging.				
I have increased my knowledge about entrepreneurial ecosystems.				
The content presented is replicable in my community.				
New connections at CEC can help me bring these practices to my community.				
This conference was a good value.				

Do you have any other feedback about the conference?

Engaging Hard-to-Reach Audiences through Internal Interdisciplinary and External Diverse Collaborations

ERIN M. GARRETT[1] AND ASHLEY J. BELLE[1]

AUTHORS: [1]University of Illinois Extension.

Abstract. Through a combination of internal and external collaborations, consumer-based energy education designed for hard-to-reach audiences was successfully delivered statewide by an interdisciplinary Extension team. Program participants representing rural residents, senior citizens, and low-income audiences demonstrated improvements in knowledge and increased intention to change their home electricity usage behaviors. This outreach work can serve as a model for other Extension services to combine interdisciplinary teams with community partnerships to reach underserved audiences statewide.

INTRODUCTION

Extension Services are renowned for their ability to disseminate research-based information through diverse, developed networks. Leveraging this network, academic professionals at University of Illinois Extension formed the Smart Grid Outreach Team to develop and deliver consumer education about electric smart grid upgrades following the rollout of smart meter installation throughout Illinois.

Our team launched energy education programs in 2017 that engaged rural, low-income, and senior audiences because many suffer from a substantial energy burden (energy costs exceeding 6% of household income). Low-income families in Illinois spend approximately 13% of their income on energy costs (Elevate Energy, 2017); 50% of these households also experience energy insecurity, the inability of a household to meet its basic energy needs (U.S. Energy Information Administration, 2018). About 20% of households with residents aged 60 or older also experience energy insecurity (U.S. Energy Information Administration, 2018). However, implementing energy efficiency upgrades can result in 15% to 30% energy savings for a household (Elevate Energy, 2017).

In this report, we describe our program design, how we developed diverse collaborations, and how the Smart Grid Outreach Team has increased participant knowledge and intent to change energy behaviors. Our outreach work can serve as a model for other Extension services to combine interdisciplinary teams with community partnerships to engage underserved audiences in making energy efficiency upgrades, especially as smart grid upgrades are occurring nationwide.

PROGRAM DESIGN

Coupling research and evidence-based practices, our team developed a diverse array of energy programs to reach our intended audiences. Over four years, we received over $900,000 in grant funding from the Illinois Science and Energy Innovation Foundation. To reach diverse underserved communities statewide, we adapted educational content to include smart grid bingo at senior centers, Lunch-N-Learn programs at Meals on Wheels, interactive electric grid models for youth, and demonstrations of smart technology and simple residential energy efficiency upgrades. We also reached 100,000 Illinoisans with asynchronous education via radio, television, podcasts, social media, and information fairs. We designed our programming to help consumers understand how their electricity reaches them, why a smarter system helps manage energy use, and how to improve home energy efficiency, with the goal of helping them save money and reduce demand on the electric grid.

DIVERSE COLLABORATIONS

Collaboration—both internally across Extension program areas and externally through community partnerships—maximizes available resources, expands access to specific audiences, and boosts public awareness of projects (Seevers & Stair, 2015; Stearns, 2018). Our team combined educators from two Extension teams: our energy and environmental stewardship educators developed and delivered educational programming using their expertise on energy efficiency, the smart grid, and renewable energy; our community and economic development educators connected energy educators to municipal audiences, delivered programming to senior citizen and local government audiences, managed the evaluation tool, and tracked where programming occurs.

Externally, we developed statewide collaborations with three other Illinois Science and Energy Innovation Foundation grant-funded organizations, whom we connected with at grantor networking meetings. Collaborating with the Smart Grid for Schools program (now called Smart Grid for All) at Illinois State University (a non-land-grant university), Seniors Independent Living Collaborative, and the Citizens Utility Board provided educational materials and expertise to improve our programming (Table 1). Recognizing youth as future decision makers, we used the Smart Grid for Schools hands-on miniature neighborhoods to teach youth how electricity is generated and flows through the grid. We also used these miniature neighborhoods to conduct power outage simulations with and without a smart grid (Figure 1).

Partnering with local groups already serving hard-to-reach audiences allowed our team to reach individuals who may not have attended a traditional Extension program (Bairstow et al., 2002). Our team developed the "Save Energy, Save Money" program to assist low-income participants to achieve financial savings through energy efficiency practices. Through partnership with C.E.F.S. Economic Opportunity Corporation, we delivered this interactive experience to participants of the Low Income Home Energy Assistance program, the Percentage of Income Payment Plan program, Meals on Wheels, and Head Start in central Illinois.

PROGRAM IMPACT

From May 2017 through December 2020, we delivered 175 programs statewide, reaching 4,660 participants through direct educational programming (Figure 2). We reached 3,620 adult participants and 1,040 youth participants.

Our partnerships with C.E.F.S. and Ameren Illinois, one of Illinois' major electric utilities, provided electric bill credits, home energy conservation kits, and residential energy efficiency upgrades to over 125 households (Figure 3). Combining incentives with educational intervention has been shown to shift consumer behavior to reducing residential energy usage (Kirby et al., 2015).

To evaluate program impact, we developed and distributed an IRB-approved retrospective evaluation to adult participants. The evaluation surveyed participants' knowledge of smart meters, likelihood to improve their home's energy efficiency and tell others about smart meters, and their current enrollment and/or interest in cost savings programs. Using a Likert scale of 1 (*very low/not at all likely*) to 5 (*very high/extremely likely*), participants ranked their knowledge and likelihood before and after attending the program. Evaluations without a paired response (before and after) were excluded.

Analysis of the evaluation results using paired t-tests assuming equal variances found that after attending the program, there was a statistically significant increase in participants' knowledge of smart meters and likelihood to improve home energy efficiency (Figure 4a). Additionally, there was a statistically significant increase in participants' likelihood to enroll in energy cost savings programs such as Peak Time Rewards and Power Smart Pricing (Figure 4b).

In 2019, University of Illinois Extension recognized our team's outreach work with an Interdisciplinary State Team Excellence Award and Ameren recognized our partnership with C.E.F.S. to increase energy efficiency engagement in underserved communities with an Energy Innovator Award.

Table 1. Summary of Internal and External Partners for Smart Grid Outreach

Type of collaboration	Local or statewide effort	Partners involved	Description of partnership
Internal	Statewide	- Five energy and environmental stewardship Extension educators - Three community and economic development Extension educators	Two University of Illinois Extension teams brought separate expertise and audiences to develop the Smart Grid Outreach Team.
External	Statewide	Illinois State University (ISU), Smart Grid for All (Schools)	ISU loaned the Extension team educational electricity kits for usage in local programs.
External	Statewide	Seniors Independent Living Collaborative (SILC)	SILC provided training on the benefits of smart technology for senior audiences and provided educational materials for the Extension team to use.
External	Statewide	Citizens Utility Board (CUB)	CUB served as a program delivery partner and a source of expertise to help answer Extension questions and provide educational materials.
External	Local	C.E.F.S. Economic Opportunity Corporation	C.E.F.S. provided home energy conservation kits, residential energy efficiency upgrades, and access to local low-income audiences, including participants of the Low Income Home Energy Assistance program, the Percentage of Income Payment Plan program, Meals on Wheels, and Head Start.
Internal	Local	4-H	4-H connected the Extension team with youth audiences for in-school and after-school workshops as well as conducted Welcome to the Real World simulations for eighth graders learning money management.

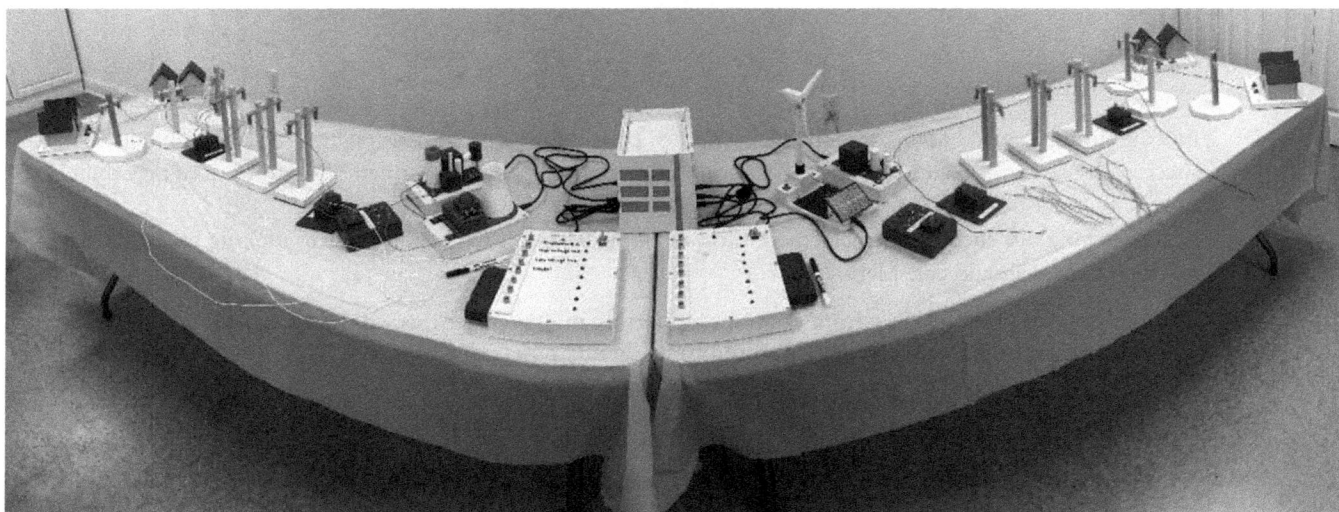

Figure 1. Smart grid for schools kit used for developing hands-on learning activities.

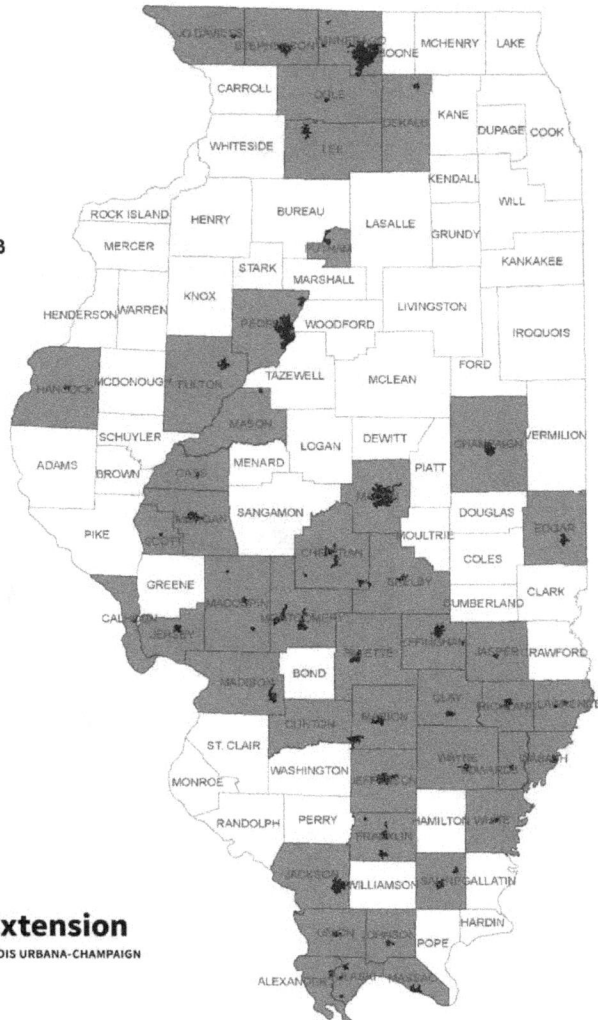

Figure 2. Map of counties and the communities within them receiving direct education from the smart grid outreach team.

CONCLUSION

By combining internal and external collaborations, programming designed for hard-to-reach audiences can be successfully delivered statewide, improving knowledge and increasing consumers' intentions to change their home electricity usage behaviors.

Following our successful engagement model (Figure 5), other Extension services looking to engage hard-to-reach audiences should consider:

- applying for grant funding from grantors supporting multiple agencies within a state and taking advantage of established networking opportunities to identify potential partners;

- finding multi-disciplinary collaborations within Extension to bolster their programming and introduce them to new audiences;

- leveraging Extension's ability to disseminate research-based information through extensive, diverse networks to find new community-based and statewide collaborators;

- synthesizing partner resources into a cohesive core message to be shared broadly; and

- integrating the core message into programming adapted to support local audience needs.

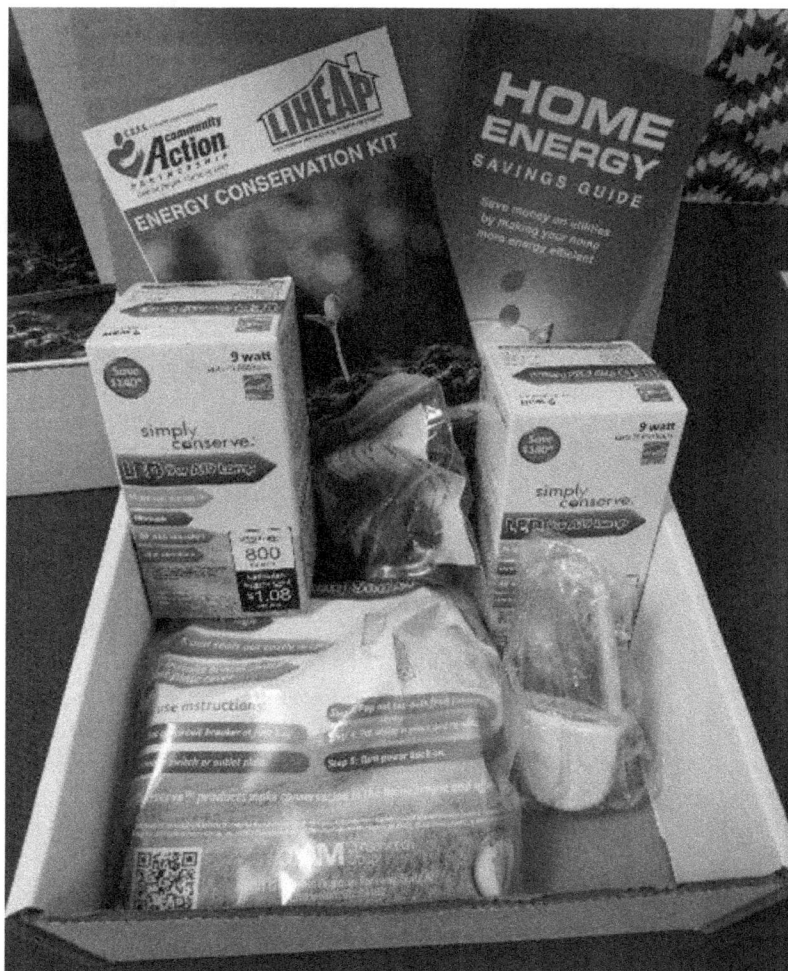

Figure 3. Home energy conservation kit. Adult program participants received an energy conservation kit containing LED light bulbs, an LED nightlight, an air filter whistle, and foam-insulating gaskets.

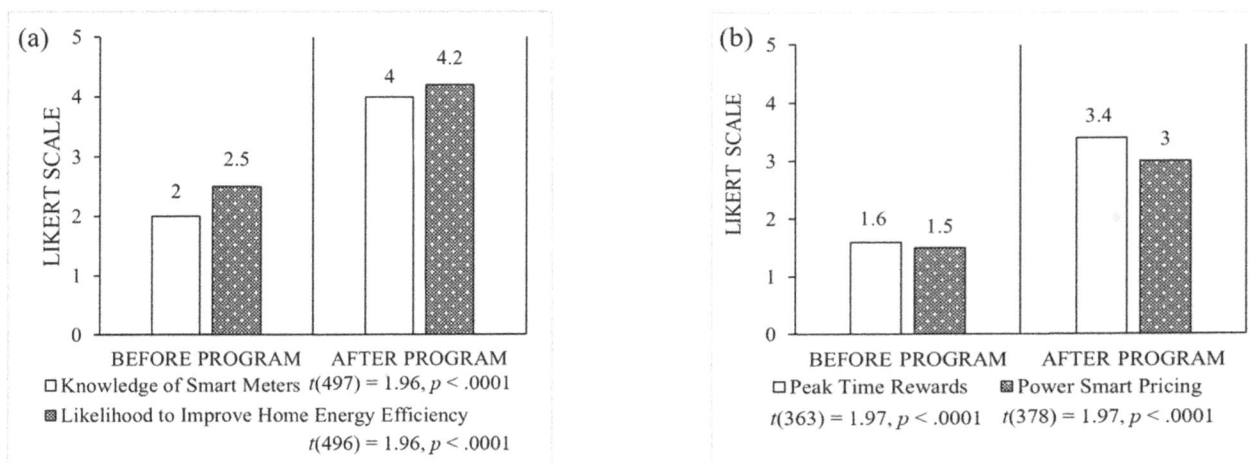

Figure 4. Retrospective evaluation analysis. Graph a shows participants' knowledge of smart meters and their likelihood to improve home energy efficiency. Graph b shows participants' likelihood to enroll in energy cost savings programs.

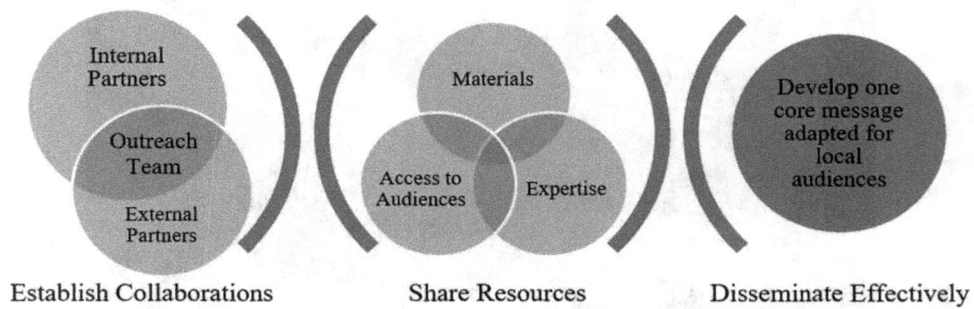

Figure 5. Model for engaging hard-to-reach audiences through internal and external collaborations.

REFERENCES

Bairstow, R., Berry, H., & Driscoll, D. M. (2002). Tips for teaching non-traditional audiences. *Journal of Extension*, *40*(6). https://archives.joe.org/joe/2002december/tt1.php

Elevate Energy. (2017). *Fact sheet: Energy burden in Illinois.* Elevate Energy. https://www.elevateenergy.org/document/fact-sheet-energy-burden-il/

Kirby, S. D., Guin, A., & Langham, L. (2015). Energy education incentives: Evaluating the impact of consumer energy kits. *Journal of Extension*, *53*(1). https://tigerprints.clemson.edu/joe/vol53/iss1/16/

Seevers, B., & Stair, K. (2015). Exploring community partnerships in agricultural and extension education. *Journal of Extension*, *53*(3). https://tigerprints.clemson.edu/joe/vol53/iss3/15/

Stearns, S. (2018). Developing internal partnerships to enhance a local foods campaign. *Journal of Extension*, *56*(4). https://tigerprints.clemson.edu/joe/vol56/iss4/8/

U.S. Energy Information Administration. (2018). *Residential energy consumption survey 2015* [Data set]. Residential Energy Consumption Survey. https://www.eia.gov/consumption/residential/data/2015/

JOURNAL OF
Extension

Ideas at Work

Volume 60, Issue 2, 2022

Expanding Effective Behavioral Health Literacy Programs to Address Farm Stress

CHERYL L. ESCHBACH[1], COURTNEY CUTHBERTSON[2], GWYN SHELLE[1], AND RONALD O. BATES[1]

AUTHORS: [1]Michigan State University Extension. [2]University of Illinois.

Abstract. Attention to stress and mental health among agricultural producers has increased over recent years, and Cooperative Extension has been active in offering educational workshops and resources to agricultural audiences. This article describes the process and effectiveness of expanding two Michigan State University Extension farm stress management programs to Cooperative Extension in other states through a national Farm Stress Management Summit. The two-day training Summit provided deeper knowledge about farm stress issues and prepared Extension professionals to offer behavioral health programs in their own communities and respective states. Evaluation findings highlight effective aspects of the Summit and next steps.

INTRODUCTION

Cooperative Extension has a history of providing research-based information to communities to enhance their wellbeing and a mission to provide resources and education to meet community needs. In that spirit, Cooperative Extension has been responsive to concerns about mental health among agricultural producers (Cuthbertson et al., 2020; Inwood et al., 2019; McMoran et al., 2019; Rudolphi & Barnes, 2019), who often face greater rates of stress, depression, and suicide than the general population (Hagan et al., 2019). The purpose of this implementation project was to train Cooperative Extension professionals in two behavioral health literacy programs that address farm stress to enable them to offer the programs in communities in their respective states. Behavioral health literacy programs focus on improving knowledge about signs and symptoms of mental health and substance use, strategies for communicating with someone who is struggling, and skills to encourage help-seeking (Jorm, 2012).

BACKGROUND ON FARM STRESS MANAGEMENT PROGRAMS

In response to increasing concern from agricultural community members and industry leaders, Michigan State University (MSU) Extension created two educational programs about farm stress to share research-based information about farmers' mental health. *Communicating with Farmers under Stress* (CFS) was created for people who work with and/or support farmers and can be delivered in two to four hours. *Weathering the Storm: How to Cultivate a Productive Mindset* (WTS) was designed for farmers and their families and can be delivered in 60 to 90 minutes. Both programs include topics such as unique stressors in agriculture, how to communicate with distressed producers, active listening and communication skills, tips on referring people to applicable resources, how to handle situations when producers might be considering suicide, background about agricultural economic trends, the impacts and signs of stress, and signs of suicide. The WTS program also includes stress identification and management techniques. From 2016 to 2019, the two farm stress programs were offered to 1,250 people with evaluations indicating programs were effective (Cuthbertson et al., 2021).

EXPANDING IMPLEMENTATION STRATEGIES TO ADDRESS FARM STRESS

The Farm Stress Management Summit, a two-day training, was held by MSU Extension in January 2019 for Cooperative Extension professionals from other states to be trained in the MSU Extension farm stress programs. Two main objectives were to:

1. Provide in-depth context about farm stress

2. Prepare participants to implement the farm stress programs in their communities.

Mental Health First Aid (MHFA) training was a prerequisite. MHFA is an evidence-based training from the National Council for Mental Wellbeing that teaches participants how to recognize the signs and symptoms of a mental health crisis and how to help someone who may be experiencing one (Hadlaczky et al., 2014; Kitchener & Jorm, 2002).

The Summit lasted two full days. The first day included professional development on topics related to farm stress. See Table 1 for the training agenda. The shared professional development provided current, in-depth information for Summit attendees and generated a cross-state conversation on agricultural markets and farm stressors. On the second day, participants received both farm stress programs, facilitated by MSU Extension professionals; participants then practiced presenting part of one program in small groups with feedback from MSU Extension educators.

After completing training, participants were eligible to offer the programs in their states. Facilitators received access to full training materials, including scripted PowerPoint presentations, instructor guides for facilitation and implementation, and 12 handouts. Materials were designed to be co-branded with universities of new facilitators. MSU Extension included a standardized program evaluation tool and collects data from participating states to assess outcomes from trainings.

Participants included 99 people from 23 states (see Table 2). Nearly one-third (29%) had worked with farm families for over 20 years, and 43% had worked for Cooperative Extension for over 10 years. Half of participants were either current farmers or had farmed in the past.

EFFECTIVENESS

Summit participants completed evaluations about the effectiveness of the training, improvements in their knowledge, and plans for future action. Evaluations were collected at the Summit for day 1 (n=74) and day 2 (n=62). Table 3 shows evaluation findings from both surveys by outcome.

In addition to quantitative data, researchers collected comments on what participants saw as the most valuable parts of the training. Comments reflected that participants learned a great deal from the Summit, and included statements such as:

- "I learned from the conference and talking to other Extensions, ways to deal with the farmers in distress."

- "The [Congressman's] account taught me a lot of lessons about the agrarian imperative.... Knowing the psychological underpinnings for why so many farmers make a seemingly irrational decision will be helpful in the process of connecting with them."

- "A deeper understanding of the background issues affecting farmers will provide great context on delivering farm stress programming."

- "The financial information will be helpful when reaching audiences that respond better to empirical reasoning."

- "I think the discussion of how to have conversations is especially helpful."

Table 1. Agenda for Two-Day Farm Stress Management Training Summit

Type of Presenters	Session Topics
Extension Professionals (e.g., Specialists, Educators, Administrative Leadership)	Agricultural economics, farm family dynamics, mental health and how to help distressed farmers, a role-playing activity with farm situations for discussions.
Elected Officials/Farmer	A state legislator's personal account of farm stress and agribusiness story.
Practitioner/Farmer	Dr. Michael Rosmann, an Iowa farmer and trained Psychologist, presented on the agrarian imperative and its importance to farmer identity.
University Professors	Overview of agricultural market data and economic trends for different commodities; decision-making tools for financial planning.

Table 2. Participating States and Universities at MSU Extension Farm Stress Management Training Summit

Cooperative Extension Region	States	Represented Universities and Colleges
Northeast	DE, MD, NY, PA	University of Delaware, University of Maryland, Cornell University & NY Farm Net, Pennsylvania State University
North Central	IA, IL, KS, MI, MN, NE, OH, SD, WI	University of Iowa, University of Illinois, Kansas State University, Michigan State University, Central Lakes College, University of Nebraska, The Ohio State University, South Dakota State University, University of Wisconsin
Southern	MS, NC, VA	Mississippi State University, East Carolina University, North Carolina State University, Virginia Polytechnic and State University
Western	ID, HI, MT, OK, OR, WA	University of Idaho, University of Hawaii, Montana State University, Oklahoma State University, Oregon State University, Washington State University
1890	GA	Fort Valley State University

Note. Although the programs were developed for delivery through Cooperative Extension, two Summit participants were not from land-grant institutions.

Table 3. Evaluation Findings for the MSU Extension Farm Stress Management Training Summit

Outcome	Indicators
Increased confidence to offer farm stress programs based on Summit training	• 96% reported the Mental Health First Aid training prerequisite of the Summit training raised their confidence to offer the farm stress management programs • 97% reported the workshop demonstration portion raised their confidence to offer the farm stress management programs • 86% reported the teaching practice experience portion raised their confidence to offer the farm stress management programs
Likelihood of implementing programs in home state	• 95% reported they were likely to implement the "Communicating with Farmers Under Stress" program • 97% reported they were likely to implement the "Weathering the Storm: How to Cultivate a Productive Mindset" program
Improved knowledge about various aspects of farm stress	• 99% improved their knowledge of agriculture markets and economic trends • 95% improved their knowledge of the agrarian imperative and the importance of farmer identity • 88% of attendees gained knowledge on unique challenges farmers face; others reported they knew this information already • 95% improved their knowledge of how to help farmers in distress with farm business tools
Improved knowledge of strategies to work with farm-related audiences, and plans for action after training	• 83% learned new strategies to work with agribusiness in their local communities • 77% anticipated multi-state collaborations or working together with people from other states that they met at the Summit • 95% reported that the summit provided an opportunity to build a support network for farmers and their families

NEXT STEPS

For accessibility and sustainability, and due to increased demand following the Summit, MSU Extension created an online train-the-trainer program for additional states and educators to be able to offer CFS and WTS. This online train-the-trainer program is offered through the learning management system Desire2Learn (D2L), and the course is asynchronous, allowing participants to complete it during the times that work best for them. This online train-the-trainer program is a combination of recorded presentation slides, interactive activities, and online discussion boards. Completion grants the new facilitator access to an online learning community for individuals trained in the farm stress programs to collaborate and access materials. Upon completion of the course, newly trained facilitators receive a certificate and access to the training materials so they can offer both behavioral health literacy programs in their communities. The online platform is beneficial, because it allows immediate access to updated materials, such as the 20-minute version of WTS, called *Mending the Stress Fence*. Using online training platforms helps keep Extension professionals connected across state lines, enabling collaborative program efforts, sharing of best practices, and constructive problem-solving. As of 2022, MSU Extension farm stress programs are implemented in-person or online in 23 states, indicating that the implementation of train-the-trainer summit was successful. Additionally, evaluation data from the implementation of farm stress programs has provided evidence for effectiveness with audiences (Cuthbertson et al., 2021).

REFERENCES

Cuthbertson, C., Brennan, A., Shutske, J., Zierl, L., Bjornestad, A., Macy, K., Schallhorn, P., Shelle, G. Dellifield, J., Leatherman, J., Lin, E., & Skidmore, M. (2020). Developing and implementing farm stress training to address agricultural producer mental health. *Health Promotion Practice 23*(1) 8-10. https://doi.org/10.1177/1524839920931849

Cuthbertson, C., Eschbach, C., & Shelle, G. (2021). Addressing farm stress through Extension mental health literacy programs. *Journal of Agromedicine, 27*(2), 124-131. https://doi.org/10.1080/1059924X.2021.1950590

Hadlaczky, G., Hökby, S., Mkrtchian, A., Carli, V., & Wasserman, D. (2014). Mental Health First Aid is an effective public health intervention for improving knowledge, attitudes, and behaviour: A meta-analysis. *International Review of Psychiatry 26*(40), 467-475. https://doi.org/10.3109/09540261.2014.924910

Hagen, B., Albright, A., Sargeant, J., Winder, C. B., Harper, S. L., O'Sullivan, T. L., & Jones-Bitton, A. (2019). Research trends in farmers' mental health: A scoping review of mental health outcomes and interventions among farming populations worldwide. *PloS one, 14*(12), e0225661. https://doi.org/10.1371/journal.pone.0225661

Inwood, S., Becot, F., Bjornestad, A., Henning-Smith, C., & Alberth, A. (2019). Responding to crisis: Farmer mental health programs in the Extension North Central Region. *Journal of Extension, 57*(6). https://tigerprints.clemson.edu/joe/vol57/iss6/20/

Jorm, A. F. (2012). Mental health literacy: Empowering the community to take action for better mental health. *American Psychologist, 67*(3), 231-243. https://doi.org/10.1037/a0025957

Kitchener, B. A. & Jorm, A. F. (2002). Mental Health First Aid training for the public evaluation of effects on knowledge, attitudes and helping behavior. *BMC Psychiatry, 2*(10). https://doi.org/10.1186/1471-244X-2-10

McMoran, D., Seymour, K., Bachtel, S., & Thorstenson, S. (2019). Creating a suicide prevention program for farmers and farmworkers. *Journal of Extension, 57*(6). https://tigerprints.clemson.edu/joe/vol57/iss6/11/

Rudolphi, J., & Barnes, K. L. (2019). Farmers' mental health: Perceptions from a farm show. *Journal of Agromedicine, 25*(1), 147-152. https://doi.org/10.1080/1059924X.2019.1674230

JOURNAL OF
Extension

Ideas at Work

Volume 60, Issue 2, 2022

A New Lens: Using the Policy, Systems, and Environmental Framework to Guide Community Development

CAROLINE BACKMAN[1], CLEA ROME[2], LAURA RYSER[1], REBECCA SERO, AND DEBRA HANSEN[1]

AUTHORS: [1]Washington State University Extension.

Abstract. Extension is uniquely positioned to deliver data-driven solutions to complex community issues with University applied research, particularly through crises like COVID-19. Applying the Policy, Systems and Environmental (PSE) framework to community development is an effective, innovative approach in guiding Extension leaders to create, document, and share long-term transformative change on challenging issues with stakeholders. Beyond the public health sector, applying a PSE approach to community development provides leverage points for population-level benefits across sectors. This article describes current public health approaches, methodologies, and how the PSE framework translates to other programs with four examples of high-impact, systems level Extension projects.

INTRODUCTION

This article demonstrates that beyond the public health sector, the Policy, Systems, and Environmental (PSE) approach is a useful framework to guide the work of community development leaders. By taking a PSE approach that focuses on addressing policy implications, organizational structures and systems, and the physical environment, Extension can contribute to building community resilience and addressing complex community problems—from broadband access and rural economic development to climate change adaptation. As Franz (2011) states, "Embracing the public value of Extension education means repositioning the way we describe our work from what clients learn and do to what economic, environmental, and social conditions change" (p. 3).

POLICY, SYSTEMS, AND ENVIRONMENT DEFINED

The Policy, Systems and Environmental change framework emerged in the last decade as a way to approach difficult and layered public health problems including obesity, diabetes, and cancer. *Policy* refers to the written statement of an organizational position, decision, or action. *Systems* refer to the unwritten, ongoing, and often qualitative organizational changes or decisions. *Environment* is defined as the visible changes to surroundings, whether economic, social, normal, or physical (Better Living for Texans, 2017).

Rather than focusing solely on individual choices, the PSE approach acknowledges the "interrelated, dynamic, and adaptive factors" (Lyn et al., 2013 p. S24) that influence an individual's health. Therefore, public health practitioners are addressing individual's choices as well as the broader landscape that influences personal behavior—including the policies, systems, and environments that shape those behaviors (Leeman et al., 2015). Table 1 illustrates this comparison. Within Extension, those working in the field of nutrition and health are increasingly incorporating a PSE approach in their work (Kennedy et al., 2020; Schroeder et al., 2018; Sneed et al., 2020).

UTILIZING A PSE APPROACH IN COMMUNITY DEVELOPMENT

Frequently, community development Extension work is rooted in affecting change at a policy, systems, and environment level. Beyond direct educational methods, Extension work takes place within coalitions and councils made up of agencies, organizations, and other stakeholders that work together to address complex community-level issues (Rÿser et al., 2020). Extension professionals often play a role in assisting coalitions as they work through a shared process to identify the problem, set shared goals, and execute action steps. Extension's role increases capacity with skills like facilitation, applied research, fund development, and evaluation (Smathers & Lobb, 2015). This expanded capacity enables communities to better respond to challenges, such as coordinating a response to food insecurity during the COVID-19 pandemic.

Community development Extension professionals are uniquely situated to do effective PSE work involving data-driven activity planning, community asset mapping, and the creation of goals alongside stakeholders (Rÿser et al., 2020). Described as "a portfolio of strategies at multiple levels (e.g., individual, family, community) across multiple sectors (e.g., school, worksite, neighborhood) following some variant of the socioecological model" (Cheadle et al., 2016, p. 349), PSE complements Extension's emerging, innovative work. The PSE framework creates long-term change while moving from programs to projects (e. g., from offering direct education to building coalitions that change policies, systems, and/or environments). To facilitate a process for desirable change outcomes for community development, the PSE framework (Lyn et al., 2013) outlines seven key activities:

1. Assess the social and political environment.

2. Engage, educate, and collaborate with key stakeholders.

3. Identify and frame the problem.

4. Utilize available evidence.

5. Conduct research to identify needed data.

6. Identify PSE solutions.

7. Build support and political will.

For example, WSU Extension in Clallam County saw increased wild carrot invasion and heard concerns by the county road department in assessing and controlling spread (*assessed the environment*). Extension engaged Master Gardeners to collaborate with the county road department to identify where there were infestations and how to control the spread (*engaged stakeholders, identified the problem*). Master Gardeners recorded and shared data on wild carrot infestations on county roadsides and advised on timing and location of maintenance work to reduce spread (*utilized evidence and identified needed data*). This collaboration improved county-wide sustainable weed management policies (*identified PSE solutions)*, while building political will that led to new collaborative native roadside plantings (*built political will*). Table 2 demonstrates this and other community development projects completed by Extension faculty and explains how that work relates to the PSE framework.

Providing decision makers with data for effective policy decisions (Dodd & Abdalla, 2004; Rome & Lucero, 2019) complements Extension's role in community development, as does coalition building that brings together partners to identify and work on problems (Harder, 2019; Koonce et al., 2016). Extension faculty and staff also secure funding to create improved infrastructure for our communities, resulting in long-term impacts to our built environment (Sandkamp, 2014).

Table 1. Moving Toward a PSE Approach

Setting	Conventional Approach	PSE Approach
School	Teach a lesson on eating more fruit and vegetables.	Make fruits and vegetables appealing and easy to access in school cafeterias.
Workplace	Offer an employee health fair once a year.	Coordinate weekly CSA produce deliveries to the workplace.
Community	Arrange a one-time fun run for disease awareness.	Improve pedestrian and bike access on local roads and sidewalks.

Source. (Comprehensive Cancer Control National Partnership, 2015).

Table 2. Examples of Extension Community Development Creating Policy, Systems, and Environment Change

Project	Policy	Systems	Environment
Integrated Weed Management[a]	Master Gardeners collected information on the invasion of wild carrot, and in the process, shaped county-level sustainable management policies that address the larger problem of roadside weeds.	The county road department now consults annually with Master Gardeners and noxious weed control staff on appropriate timing and locations for their maintenance work to minimize the spread of weeds through mowing activities.	The Master Gardeners and noxious weed coordinator have established several plots of native roadside plantings as a demonstration of sustainable roadside management.
Marine Resources Committee (MRC)[b]	The Marine Resource Committee worked with coastal non-profits, tribes and agencies to seek dedicated funds for restoration in underserved areas of coastal Washington. These funds are now allocated by the state legislature specifically for Washington coast restoration and resiliency and managed as a state grant program.	Historically, hands-on science STEM experiences were limited by transportation and aging equipment. The MRC's additional funding to school STEM field trips and equipment over the last ten years enable consistent hands-on education in the high school on stream restoration and fish biology.	The Marine Resource Committee partnered with the school district to create a value-added food and fish processing facility. The committee secured funding for the facility, provided university food safety training to fishers, and completed the 1440 sq. ft. commercial food processing facility for school and fishers' use.
Community Food Systems[c]	The Kitsap Food Policy Council worked with community partners to write new food policies for the 2016 Comprehensive Plan update.	The Healthy Eating, Active Living collective impact coalition of 29 organizations and agencies aligned strategies to improve community-level health outcomes through increased access to opportunities for activity and fresh food.	The Kitsap Fresh cooperative operates a weekly online farmers market that connects food buyers with businesses selling locally produced and processed foods and distributes food throughout the county.
Broadband Action Team (BAT)[d]	The BAT has directly influenced major legislation recently introduced by Senator Murray—the Digital Equity Act of 2019—that would look to provide over $1 billion in grants over five years to stand-up and support BAT-like organizations.	The BAT provides a "one-stop shop" forum and website for all stakeholders to share resources, issues, solutions and collaborative projects.	The City of Chewelah held a BAT community meeting, resulting in having an Internet Service Provider lay fiber.

Note. [a](Rome & Lucero, 2019), [b](Backman, 2020), [c](Kitsap County Department of Community Development, 2016), [d](Rÿser et al., 2020).

THE IMPORTANCE OF ADOPTING THIS FRAMEWORK

With the advent of the information age, Extension professionals need to adapt services to remain relevant (King & Boehlje, 2000). With increased competition, Extension as a source for information alone is increasingly irrelevant. For example, King (2018) found that in Oregon, between 1986 and 2006, the number of people who said they had used Extension in the preceding year dropped from 44% to 22%. In this changing environment, Extension has struggled to find ways to help decision-makers understand the public value of Extension work (McGrath et al., 2007).

With PSE widely accepted as an understandable and effective framework (Honeycutt et al., 2015), the authors propose that it can be an important platform for communicating the critical, unique, and sometimes nebulous and unseen role we play in community development work. PSE can give new relevance to the important work that Extension professionals do in communities across the country. Extension is uniquely positioned to create PSE change because of their intrinsic collaborative skillsets and abilities to leverage strategic partners and media (Jouridine & Green, 2001; Surls & Hayden-Smith, 2013).

PSE AND COVID-19

The COVID-19 pandemic has illustrated vulnerability in many policies and systems, affecting almost every aspect of the environments in which we live and work. Extension professionals working in all fields have been adapting, pivoting, and responding to this new challenge in diverse ways, many relying on the foundations of PSE work laid prior to the pandemic. The network of community relationships and coalitions built by Extension agents has become a powerful way to exchange information, strategize effective responses, and stave off isolation in this newly fragmented environment (Nash et al., 2020; Washburn et al., 2020).

CONCLUSION

The authors propose that the PSE approach is an effective framework with which to guide the work of community development-focused faculty and staff in Extension. In addition to creating community-level impacts addressing complex issues, the PSE framework can effectively communicate the work of Extension to audiences both internally (at the University) and externally (to stakeholders and funders). Moving forward, Extension is seeking to be relevant and effective in addressing 21st century problems. Leading the charge in facilitating PSE change in communities, in addition to maintaining the traditional direct education role, positions Extension professionals at the forefront of helping communities address these emerging challenges.

REFERENCES

Backman, C. (2020, March 17). Invited Remote Presentation on Rural Food Processing. Association of Public and Land Grant Universities, Commission on Economic and Community Engagement.

Better Living for Texans (2017, November 13). *Policy, Systems, & Environmental (PSE) Definitions and Examples.* Texas A&M AgriLife Extension. https://blt.tamu.edu/files/2017/11/Policy-Systems-Environment-Change-Definitions-Examples.pdf

Cheadle, A., Cromp, D., Krieger, J. W., Chan, N., McNees, M., Ross-Viles, S., Kellogg, R., Rahimian, A. & MacDougall, E. (2016). Promoting policy, systems, and environment change to prevent chronic disease: Lessons learned from the King county communities putting prevention to work initiative. *Journal of Public Health Management and Practice, 22*(4), 348-359. www.doi.org/10.1097/PHH.0000000000000313

Comprehensive Cancer Control National Partnership. (2015, May 15). *Policies, Systems and Environmental Change Resource Guide.* https://www.acs4ccc.org/wp-content/uploads/2020/06/PSE_Resource_Guide_FINAL_05.15.15.pdf

Dodd, A., & Abdalla, C. (2004). Strengthening environmental policy education through qualitative research: Experience with Pennsylvania's Nutrient Management Act regulatory review. *Journal of Extension, 42*(5). https://archives.joe.org/joe/2004october/a2.php

Franz, N. K. (2011). Advancing the public value movement: Sustaining Extension during tough times. *Journal of Extension, 49*(2). https://tigerprints.clemson.edu/joe/vol49/iss2/11

Harder, A. (2019). Public value and partnerships: Critical components of Extension's future. *Journal of Extension, 57*(3). https://tigerprints.clemson.edu/joe/vol57/iss3/24

Honeycutt, S., Leeman, J., McCarthy, W. J., Bastani, R., Carter-Edwards, L., Clark, H., Garney, W., Gustat, J., Hites, L., Nothwehr, F., & Kegler, M. (2015). Evaluating Policy, Systems, and Environmental change interventions: Lessons learned from CDC's Prevention Research Centers. *Preventing chronic disease, 12*(10), E174. http://dx.doi.org/10.5888/pcd12.150281

Jouridine, L. A., & Green, S. D. (2001). Extending our reach: Strategic opportunities for Cooperative Extension to promote infant health through Sudden Infant Death Syndrome preventative education. *Journal of Extension, 39*(3). https://archives.joe.org/joe/2001june/a8.php

Kennedy, L., Sneed, C. T., Franck, K. L., Norman, H., Washburn, L., Jarvandi, S., & Mullins, J. (2020). Policy, Systems, and Environmental change: A planning tool for community health implementation. *Journal of Extension, 58*(4). https://tigerprints.clemson.edu/joe/vol58/iss4/5

King, D. A., & Boehlje, M. D. (2000). Extension: On the brink of extinction or distinction? *Journal of Extension, 38*(5). https://archives.joe.org/joe/2000october/comm1.php

King, D. (2018). Hey, Siri, what is the future of Extension? *Journal of Extension, 56*(5). https://tigerprints. clemson.edu/joe/vol56/iss5/18

Kitsap County Department of Community Development (2016). *Kitsap County comprehensive plan 2016-2036.* Kitsap County. https://www.kitsapgov.com/dcd/Pages/2016_Comprehensive_Plan.aspx

Koonce, J., Scarrow, A., & Palmer, L. (2016). Volunteer income tax assistance: A community coalition for financial education and asset building. *Journal of Extension, 54*(2). https://tigerprints.clemson.edu/joe/vol54/ iss2/7

Leeman, J., Myers, A. E., Ribisl, K. M., & Ammerman, A. S. (2015). Disseminating policy and environmental change interventions: Insights from obesity prevention and tobacco control. *International Journal of Behavioral Medicine, 22*(3), 301-311. www.doi.org/10.1007/s12529-014-9427-1

Lyn, R., Aytur, S., Davis, T. A., Eyler, A. A., Evenson, K. R., Chriqui, J. F., Cradock, A. L., Goins, K. V., Litt, J., & Brownson, R. C. (2013). Policy, systems, and environmental approaches for obesity prevention: A framework to inform local and state action. *Journal of Public Health Management and Practice, 19*(Supplement 1), S23–S33. https://doi.org/10.1097/PHH.0b013e3182841709

McGrath, D. M., Conway, F. D. L., & Johnson, S. (2007). The Extension hedgehog. *Journal of Extension, 45*(2). https://archives.joe.org/joe/2007april/a1.php

Nash, A., Brown, S. C., & Cascio, J. (2020). Importance of best guesses in emergency situations. *Journal of Extension, 58*(3). https://tigerprints.clemson.edu/joe/vol58/iss3/31

Rome, C., & Lucero, C. (2019). Wild carrot (Daucus carota) management in the Dungeness Valley, Washington, United States: The power of citizen scientists to leverage policy change. *Citizen Science: Theory and Practice, 4*(1), 36. http://doi.org/10.5334/cstp.201

Rÿser, L., Backman, C., Rome, C., Hansen, D., & Babine, M. (2020). *A policy, systems, and environmental framework: Advancing Extension's role amidst the COVID-19 pandemic.* Rural Connections. https://www.usu.edu/ wrdc/files/news-publications/Ryser-RC-FA-WIN-2020.pdf

Sandkamp, L. (2014). Cooking up innovation: A guide to creating a shared commercial kitchen facility. *Journal of Extension, 52*(6). https://tigerprints.clemson.edu/joe/vol52/iss6/24

Schroeder, M., Grannon, K., Shanafelt, A., & Nanney, M. S. (2018). Role of Extension in improving the school food environment. *Journal of Extension, 57*(7). https://archives.joe.org/joe/2018december/iw2.php

Smathers, C. A. & Lobb, J. M (2015). Extension professionals and community coalitions: Professional development opportunities related to leadership and Policy, System, and Environment change. *Journal of Extension, 53*(6). https://tigerprints.clemson.edu/joe/vol53/iss6/5

Sneed, C. T., Franck, K. L., Norman, H., Washburn, L., Kennedy, L. E., Jarvandi, S., & Mullins, J. (2020). Two states, one mission: Building policy, systems, and environmental change capacity of county Extension educators. *Journal of Extension, 58*(4). https://tigerprints.clemson.edu/joe/vol58/iss4/10

Surls, R.A., & Hayden-Smith, R. (2013). UC Cooperative Extension's collaborations grow with the centuries. *California Agriculture, 67*(3), 118. http://calag.ucanr.edu/archive/?type=pdf&article=ca.v067n03p118

Washburn, L., Martin, A., & Barnes, S. (2020). Activating volunteers for statewide COVID-19 pandemic response. *Journal of Extension, 58*(4). https://tigerprints.clemson.edu/joe/vol58/iss4/25

Lessons Learned Recruiting Comparison Elementary Schools for Impact Evaluation of SNAP-Ed Interventions

AMANDA LINARES[1], PHOEBE HARPAINTER[1], KAELA PLANK[1], AND GAIL WOODWARD-LOPEZ[1]

AUTHORS: [1]University of California, Division of Agriculture and Natural Resources.

Abstract. To determine the effectiveness of Supplemental Nutrition Assistance Program- Education (SNAP-Ed) nutrition and physical activity programming in elementary schools, it is necessary to recruit socioeconomically similar comparison schools not receiving SNAP-Ed programming. We developed a flexible recruitment strategy to tailor our approach to each individual school district and site. Here we discuss the lessons learned during the 10-month recruitment period, including early outreach, emphasizing participation benefits, leveraging and building relationships, and visiting sites.

INTRODUCTION

Ninety-five percent of U.S. children attend school, demonstrating the potential reach of nutrition and physical activity (Nut-PA) programming delivered in this setting (Snyder & Dillow, 2015). Evaluation of such programs, like those administered by Extension or the Supplemental Nutrition Assistance Program- Education (SNAP-Ed), is essential to ensure the investment produces desired outcomes. Despite administrator and teacher preference for Nut-PA education to occur during the school day, incorporating programming—and particularly evaluation—is challenging (Cupp et al., 2006; Frye et al., 2002; Harrell et al., 2000; Hermann et al., 2011). Use of a comparison group in evaluation is essential for attributing outcomes to the intervention (United States Department of Agriculture [USDA], 2005). However, engaging, recruiting, and retaining comparison schools not receiving programming can be difficult. In this paper, we describe lessons learned during recruitment of 36 comparison elementary schools.

DESCRIPTION OF EVALUATION

Evaluators conducted a three-year evaluation of California Local Health Department (LHD)-administered school-based SNAP-Ed programming consisting of an annual pre and post (fall and spring) Nut-PA survey of fourth/fifth grade students, supplemented with a self-administered site-level assessment of Nut-PA-related policies and practices (Rider et al., 2020). In California, SNAP-Ed consists of education (e.g., Nut-PA classes or promotional materials) and policy, systems, and environmental change approaches (e.g., nutrition standards for school foods and improved wellness policies). Comparisons receive no intervention, but agree to the evaluation activities, an estimated time commitment of six hours annually. For their participation, comparison sites receive $500 per year, along with their annual evaluation results. Due to the SNAP-Ed funding structure and grant cycle, delayed intervention or guarantee of future programming cannot be made.

COMPARISON SITE RECRUITMENT

The study objective was to determine the impact of SNAP-Ed programming on elementary students in comparison to regular practice. Therefore, we needed to recruit comparison schools with similar sociodemographic

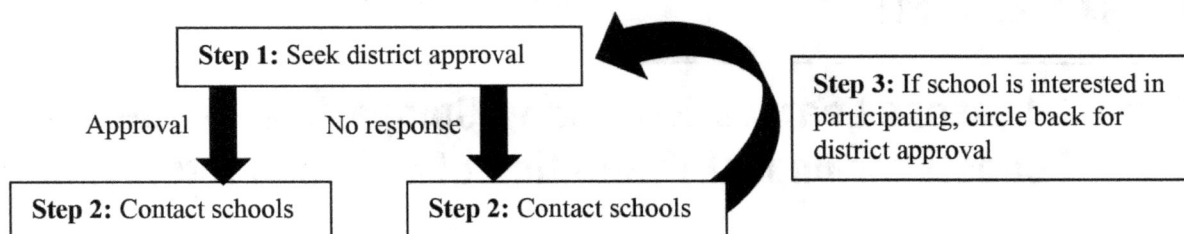

Figure 1. Steps in acquiring district approval.

characteristics that instituted programming that occurs naturally in the absence of SNAP-Ed. Inclusion/exclusion criteria for comparisons included: at least 50% of students eligible for Free/Reduced Price Meals; no current/recent SNAP-Ed programming; no externally funded Nut-PA programming. Schools whose usual practice included Nut-PA classes or gardens were not excluded, even if practice was high quality. Site-level assessments were used to control for variations in existing practices. To the extent possible, districts/schools were prioritized based on location, enrollment, and sociodemographic similarity to intervention schools. For feasibility, we prioritized those that lacked a research proposal requirement.

Flexible recruitment procedures (Figure 1) were developed so that we could effectively and consistently proceed based on the responses we received at each step in the recruitment process. Districts/schools were contacted over a 10-month period using these three steps.

RESULTS

Figure 2 outlines the recruitment process. Researchers identified 629 schools in 189 districts as potentially eligible comparisons. Based on selection criteria and available time, 71 of these districts were contacted. Of these, two districts (10 schools) were recruited by leveraging existing relationships. Of the remaining 69 districts contacted, 52 of 69 districts, managing 228 schools, provided either no response or active consent to contact principals. Of these 228 schools in districts that provided approval or no response, 71 agreed to be screened to confirm eligibility. Of the 71 schools screened, 26 were eligible, had confirmed district approval, and agreed to project commitments. A total of 36 comparison schools were selected for the three-year evaluation.

LESSONS LEARNED

Researchers identified the following lessons in recruiting comparison schools:

1. Fall and winter are ideal for recruitment, but recruiters should begin the school year prior. Time of year plays an important role in school administration decisions to participate in evaluation activities (Befort et al., 2008). Almost 90% of schools were recruited between September and February, indicating that those months are the most fruitful for recruitment. Although school administrative offices were easy to reach in May-early July before summer closure, few schools were recruited between May and August because schools were focused on wrapping up the year and principals took summer break. Fall/winter recruitment success is consistent with previous findings, as by spring, state testing and other commitments take precedence (Befort et al., 2008). Timeline in relation to evaluation start is also important. For this evaluation to measure change over time in school-year interventions, intervention and comparison pre-testing was slated to take place in fall (before intervention began) and post-testing was slated to take place in spring (after interventions are complete). Although there was interest from schools in January/February, some administrators did not want to participate until the following school year, which was too late for year one pre-testing because interventions had begun.

2. Build relationships with multiple decision makers. Because principals are key decision makers, recruitment efforts centered on reaching them. However, many schools had other administrators (like vice principals, Guidance Instructional Specialists, and school nurses) that were easier to reach and had similar latitude in decision-making. These leaders, often identified through school websites, facilitated and expedited recruitment. Furthermore, building a relationship with the school office manager and gaining his/her trust was key, because the office manager serves as a gatekeeper to the principal.

Figure 2. Steps in the recruitment process.

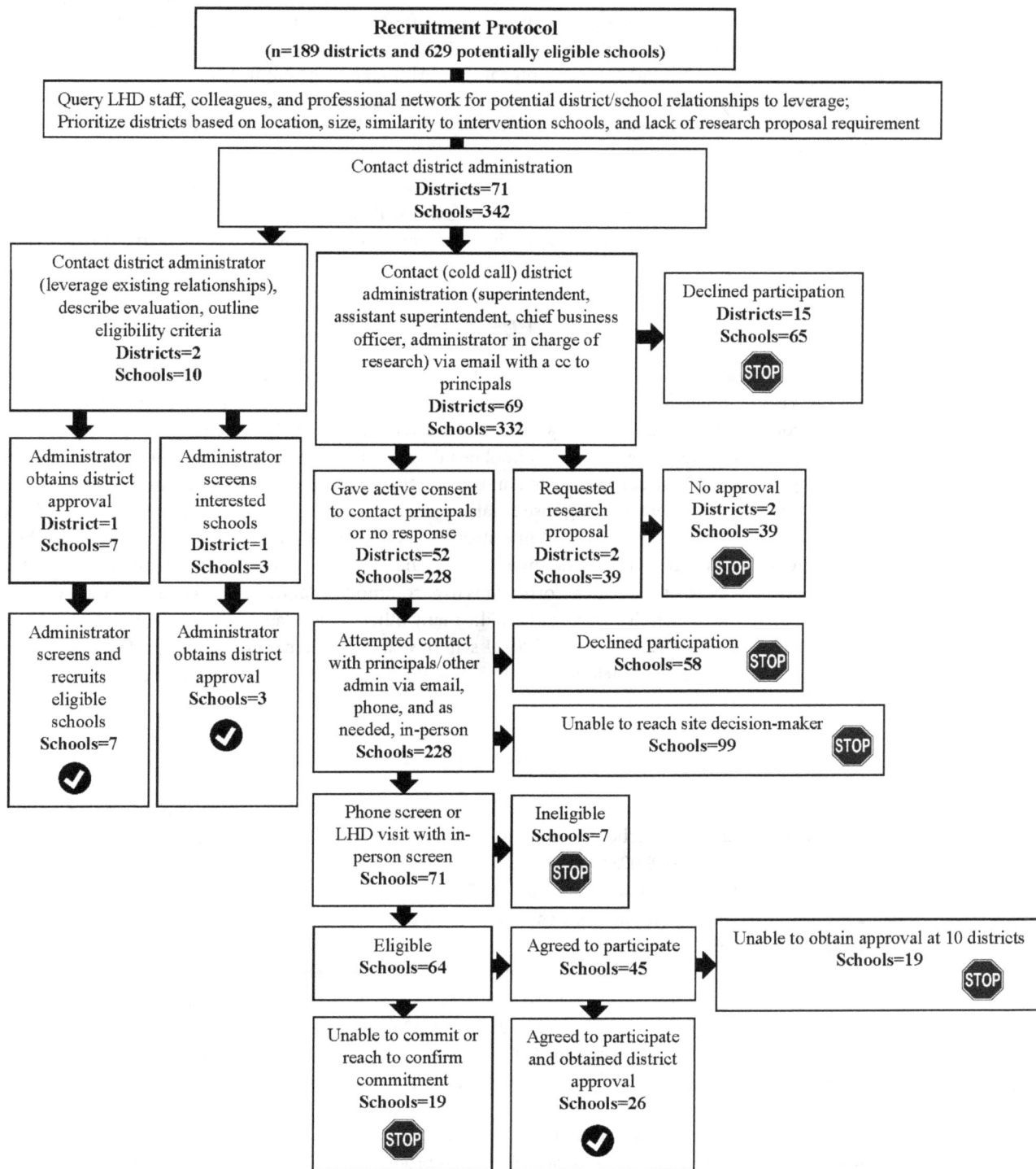

3. Clearly and succinctly emphasize benefits of participation. Practical and tangible benefits can be more important to administrators than monetary incentives (Befort et al., 2008). A few months into recruitment, materials were revised to be more concise and compelling. Materials highlighted SNAP-Ed's role in supporting school Nut-PA, and how their school's participation would benefit children throughout the state—a benefit beyond monetary incentives and access to analyzed data.

4. Consider scheduled, in-person visits. It became difficult to reach and ultimately solidify participation by phone/email with some promising sites. It was impractical for state-level recruitment staff to travel to schools across the state, so LHD staff visited prospective schools in their county as needed. Some of these visits were scheduled, while others were impromptu. Although impromptu visits afforded immediate face time, unannounced visits were not always welcome for safety reasons. Therefore, impromptu visits should only be conducted by recruiters who are at schools for other purposes and/or by those with prior relationships with administration.

5. Leverage existing relationships. Where possible, recruitment staff leveraged existing personal and professional relationships (their own and those of LHD staff) with school districts. Having an "in" provided a district-level point person who could champion the project, recruit multiple schools in a district, and facilitate approval.

CONCLUSION

The best practices outlined here provide guidance to Extension professionals evaluating school-based programs utilizing a comparison sample. Recruitment of schools presents unique and evolving challenges. Increases in time needed to fulfill educational mandates and state testing requirements means less flexibility for programming and evaluation beyond the common core subjects. School health and safety concerns make it harder for outside organizations to engage schools. These competing demands mean that recruitment processes must remain flexible and innovative. Recruiters must present a strong case regarding the benefits of participation and leverage professional networks to gain access to decision makers. While strong selection criteria for a comparison sample is the gold standard, evaluators should consider relaxing their criteria when the pool of potential comparisons is limited. In these instances, evaluators can use statistical corrections to accommodate for differences in demographics between intervention and comparison sites. This evaluation, like many others, was adapted due to COVID-19. Substantial loss of instructional time in the 2019-20 and 2020-21 school years is a potential barrier to participation in evaluation activities going forward. Nonetheless, providing Nut-PA programming and evaluating its effectiveness has arguably never been more important.

REFERENCES

Befort, C., Lynch, R., James, R. L., Carroll, S. L., Nollen, N., & Davis, A. (2008). Perceived barriers and benefits to research participation among school administrators. *Journal of School Health, 78*(11), 581-586. https://doi.org/10.1111/j.1746-1561.2008.00349.x

Cupp, P. K., Zimmerman, R. S., Massey, C. S., Howell, J. R., & Swan, R. (2006). Using community ties to facilitate school-based prevention research. *Health Promotion Practice, 7*(4), 459-466. https://doi.org/10.1177/1524839905278870

Frye, F. H., Baxter, S. D., Thompson, W. O., & Guinn, C. H. (2002). Influence of school, class, ethnicity, and gender on agreement of fourth graders to participate in a nutrition study. *Journal of School Health, 72*(3), 115–120. https://doi.org/10.1111/j.1746-1561.2002.tb06528.x

Harrell, J. S., Bradley, C., Dennis, J., Frauman, A. C., & Criswell, E. S. (2000). School-based research: Problems of access and consent. *Journal of Pediatric Nursing, 15*(1), 14–21. https://doi.org/10.1016/S0882-5963(00)80019-6

Hermann, J., Parker, S., Phelps, J., & Brown, B. (2011). What do schools want? Assessing school administrator and teacher preferences related to nutrition education program scheduling. *Journal of Extension, 49*(3), tt8. https://archives.joe.org/joe/2011june/tt8.php

Rider, C. D., Linares, A., Kao, J., Becker, C., & Woodward-Lopez, G. (2020). Assessing healthful eating and physical activity practices in places children learn. *Journal of Extension, 58*(6). Retrieved from: https://tigerprints.clemson.edu/joe/vol58/iss6/28

Snyder, T. D. & Dillow, S. A. (2015). Digest of Education Statistics 2013 (NCES 2015-011). National Center for Education Statistics, Institute of Education Sciences, U.S. Department of Education. Washington, DC: USDE. https://nces.ed.gov/pubs2015/2015011.pdf

United States Department of Agriculture, Food and Nutrition Service. (2005). *Nutrition Education: Principles of Sound Impact Evaluation.* https://fns-prod.azureedge.us/sites/default/files/EvaluationPrinciples.pdf

www.ingramcontent.com/pod-product-compliance
Lightning Source LLC
Chambersburg PA
CBHW071959220326
41599CB00034BA/7047